Are We Spiritual Machines?

Ray Kurzweil
vs.
The Critics of Strong AI

Jay Richards
George Gilder
Ray Kurzweil
John Searle
William Dembski
Michael Denton
Thomas Ray

ARE WE SPIRITUAL MACHINES?
Ray Kurzweil vs. The Critics of Strong AI
Edited by Jay W. Richards

Published in the United States
by Discovery Institute, Seattle, Washington

Library of Congress Cataloging-in-Publication Data
Richards, Jay (ed.)
    Are we spiritual machines?: Ray Kurzweil vs. the critics of strong AI /
    Jay Richards, George Gilder, Ray Kurzweil, John Searle, William Dembski, Michael Denton, Thomas Ray.
    p.   cm.
    ISBN: 0-963854-3-9
    Library of Congress Control Number: 2002101330
    1. Philosophy—United States.   2. Technology—United States.   3. Artificial intelligence. 4. Computers. I. Title.

Cover design by Joel Shoop
Layout design by Anne Hughes and Greg Piper

Books by Discovery fellows, and books published by Discovery Institute Press are available in quantity for promotional and premium use. For information on prices, terms and ordering books contact: Director of Publishing, Discovery Institute, 1402 Third Ave. Suite 400, Seattle, WA 98101 or visit the Discovery website at : www.discovery.org.

Table of Contents

Introduction

Are We Spiritual Machines?
The Beginning of a Debate

George Gilder and Jay W. Richards

11.1.04

T his volume springs from the crucible of controversy that climaxes every Gilder-Forbes *Telecosm* conference. Conducted by Gilder Publishing with *Forbes* every September in Lake Tahoe, the event brings together computer and network experts, high tech CEO's and venture capitalists, inventors and scientists. While most of the panels introduce new companies and technologies on the frontiers of the nation's economy, the closing session waxes philosophical and teleological, addressing the meanings and goals of the new machines. At *Telecosm '98* the closing topic was whether the technologies introduced in the earlier sessions might ultimately attain consciousness and usurp their creators.

George Gilder is a Senior Fellow of Discovery Institute, best-selling author of Telecosm *(New York: Free Press, 2000) and President, Gilder Technology Group. Jay Richards (editor), a philosopher and theologian, is a Senior Fellow and Program Director for Discovery Institute's Center for the Renewal of Science and Culture.*

Affirming the hypothesis of man-machine convergence is a theory called "Strong Artificial Intelligence" (AI), which argues that any computational process sufficiently capable of altering or organizing itself can produce "consciousness." The final session in 1998 centered on a forthcoming landmark in the field, a book by gifted inventor, entrepreneur, and futurist Ray Kurzweil, *The Age of Spiritual Machines: When Computers Exceed Human Intelligence* (New York: Viking, 1999).

Now available in paperback, this book was a more ambitious sequel to Kurzweil's *The Age of Intelligent Machines* (Cambridge: MIT Press, 1990), in which he made some remarkably prescient predictions about future developments in information technology. He even predicted correctly the year when a computer would first beat a Chess Master at his own game. As we all know, a special purpose IBM supercomputer named Deep Blue vanquished Gary Kasparov just seven years after he had denied that such a feat was possible. At a minimum, Kurzweil demonstrated that he understood chess more fully than Kasparov understood computing.

Kurzweil's record as a technology prophet spurred interest in his more provocative prediction that within a few decades, computers will attain a level of intelligence and consciousness both qualitatively and quantitatively beyond human capabilities. Affirming Hans Moravec's assertion that even "a genetically engineered superhuman would be just a second-rate kind of robot," he concluded that further evolution of our species will be inextricably bound to our ability to enhance our bodies and minds with integrated computer prosthetics.

Needless to say, this is an affrontal idea, and we wanted some serious intellectuals to interact on it. For that reason, we brought together a diverse panel to respond to Kurzweil, albeit all critics of strong AI, which included philosopher John Searle, biologist Michael Denton, zoologist and evolutionary algorithm theorist Tom Ray, and philosopher and mathematician William Dembski. Denton and Dembski are Fellows of Discovery Institute, which helped arrange the panel.

In the discussion, a cluster of important questions emerged: What is a person? What is a human person? What is consciousness? Will a computer of sufficient complexity become conscious? Are we essentially computers ourselves? Or are we really software stuck in increasingly obsolete, fleshy hardware? Can biological organisms be reduced to their material parts? How are we related to our technology? Is the material world all there is? What is our purpose and destiny?

Although Artificial Intelligence may seem like an esoteric topic with little relevance to anything else, in fact, many of the most important questions we face from technology to theology converge on this single subject.

This *Telecosm '98* session was so engaging, and the issues it raised so important, that it deserved a wider audience. For that reason, the Discovery Institute decided to produce this volume, with the agreement and hard work of the contributors, especially Ray Kurzweil, who is the centerpiece of attention. We don't intend for it to resolve the larger issues, but rather to initiate an important conversation as well as to expose a broader audience to a topic that primarily has been the domain of philosophers, scientists and computer geeks.

Each of these scholars brings a unique expertise and perspective to the topic, including a diversity of worldviews. A clash of worldviews evokes the commonplace metaphor of an elephant hiding in the corner. Everyone sees it but few are willing to point it out. Moreover the specialists in the field all too often focus on the trunk or tail or thickness of the skin or complexity of the DNA or nature of the ecological niche alone and remain blind to the larger or holistic dimensions of the animal. Some wonder, "Is it really God?" Others banish the thought. Kurzweil speaks of spiritual machines, yet focuses on materialist explanations. Others believe that the jungle of philosophy is full of metaphysical elephants. Most of the interesting (and pachydermic) creatures in the AI debate embody larger, unannounced, and often thick-skinned assumptions about human nature and physical nature.

We would say that Kurzweil, Searle, and Ray are philosophical "naturalists" or "materialists." They assume but don't actually say that the material world "is all there is, or ever was, or ever will be" [Carl Sagan, *Cosmos* (New York: Ballantine Books, 1993, p. 4]. While they disagree on important details, they agree that everything can or at least should be described in terms of chance and impersonal natural law without reference to any sort of transcendent intelligence or mind. To them, ideas are epiphenomena of matter.

Nevertheless, they express this perspective in different ways. Kurzweil is an intriguing and subtle advocate of Strong Artificial Intelligence. He believes that with neurological architecture, sufficient complexity, and the right combination of analog and digital processes, computers will become "spiritual" like we are. His references to spirituality might lead one to suspect that he departs from naturalism. But Kurzweil is careful with his definition. By saying computers will become spiritual, he means that they will become *conscious*. While this differs from the arid materialism of Daniel Dennett, Steven Pinker and Richard Dawkins, who treat consciousness as an illusion, the identification of the spirit with consciousness is a naturalistic stratagem.

Searle shares Kurzweil's naturalism, but not his penchant for seeing computation and consciousness on a continuum. In his essay, Searle makes use of his telling Chinese Room Argument. For Searle, no computer, no matter how souped up, will ever become conscious, because, in his words, it is not *designed* to produce consciousness. The brain, unlike a computer, is designed to produce consciousness. Any artificial hardware system that does so will need to be designed with the requisite causal powers. This appeal to teleology is an odd argument for a naturalist like Searle to make, since it seems to draw on resources he doesn't have. Nevertheless, Searle sees the difference between computation and consciousness as obvious, even if the distinction doesn't fit comfortably with his metaphysical preferences.

Thomas Ray has some specific disagreements with Kurzweil, but he doesn't object to the idea that computer technology might eventually evolve its own type of conscious intelligence if allowed to do so in an appropriate environment, such as unused portions of computers and the Internet. After all, he reasons, environment and purposeless selection evolved biological organisms into conscious intelligence on Earth. Who's to say that, given the right environment and selection pressure, our information technology won't do the same?

Denton and Dembski, in contrast, believe there's more to reality than the material world. Denton is cautious here, but he implies that biological organisms transcend physics and chemistry. While they may depend on these lower levels, they can't be reduced to them. Thus he speaks of "emergence" in a different way. While materialists use the term to mean the spontaneous and unguided evolution of complex phenomena, Denton implies that things evolve according to some sort of intelligent plan or purpose. The mechanistic paradigm forces a false reductionism of organisms to machines. Kurzweil, according to Denton, has not properly attended to the important distinctions between them.

Dembski's critique is more explicitly theistic, and, like Denton, he criticizes Kurzweil for identifying persons with machines. He is the only one explicitly to criticize Kurzweil for what he calls "tender-minded materialism." In his essay he argues that Kurzweil's materialism doesn't do justice to human persons and intelligent agents generally. Like Searle, but from a quite different perspective, he says that Kurzweil has underestimated the challenges to his project.

To avoid an unbalanced and unfair volume, in which four critics line up against one advocate, we have included Kurzweil's individual responses to his critics, and left them without editorial comment. Inspired by Thomas Jefferson, the Virginia Statute of Religious Liberty is appropriate here: "Truth is great and will prevail if left to herself. . . . She is the proper and sufficient antagonist to error, and has nothing to fear from the conflict, unless by human interposition disarmed of her natural weapons, free argument and debate."

Bill Joy's Left Turn

Kurzweil's ideas have already had a profound effect on those who have heard them. One well-known effect came from a *Telecosm* '98 conferee who missed the actual Kurzweil session. Bill Joy, co-founder of Sun Microsystems, happened to sit with Kurzweil in the lounge after the closing session, and Kurzweil briefed him on his vision for the future. Joy was deeply affected, because he knew that Kurzweil was one of the great intellects of the industry, a pioneer of computer voice recognition and vision. Coming from Kurzweil, what had previously seemed like science fiction now appeared to Joy as "a near-time possibility." As a result, Joy published his alarm in the April 2000 issue of *Wired* ("Why the future doesn't need us.") The eloquent and stirring ten-thousand word personal testament evoked more mail and comment than any previous article in the magazine (or in any other magazine in recent memory). The difference is that while Kurzweil is upbeat about the future he sees, Joy is filled with dread.

Kurzweil's argument, and now Joy's, drastically compressed and simplified, is that Moore's Law, which for almost 40 years has predicted the doubling of computer power roughly every 18 months, is not going to expire later this decade. Of course *traditional* chip manufacturing techniques will hit the quantum barrier of near-atomic linewidths. Nevertheless, Joy now believes that "because of the recent rapid and radical progress in molecular electronics—where individual atoms and molecules replace lithographically drawn transistors—and related nanoscale technologies, we should be able to meet or exceed the Moore's Law rate of progress for another thirty years." The result would be machines a million times as fast and capacious as today's personal computers and thereby "sufficient to implement the dreams of Kurzweil and Moravec," that is, intelligent robots by 2030. And "once an intelligent robot exists it is only a small step to a robot species—to an intelligent robot that can make evolved copies of itself."

Joy's nightmares do not stop with sentient machines. Intimately related are the very "nanotechnologies" that may enable the extension of Moore's Law with genetic engineering. Central to this GNR (Genetics, Nanotechnology, Robotics) trinity of techno-terror are two characteristics that could hardly be better calculated to inspire fear. The first is what Joy calls the "dematerialization" of industrial power. In the past, you needed rare resources, large nuclear plants, and huge laboratories to launch a new holocaust. In the future you will need only a computer and a few widely available materials.

The even more terrifying common thread is "self-replication." As enormous computing power is combined with "the manipulative advances of the physical sciences" and the revealed mysteries of genetics, "The replicating and evolving processes that have been confined to the natural world are about to become realms of human endeavor." New germs are self-replicating by definition. So too, Joy's "robot species." And then there is the gray goo.

Joy's journey from his Silicon Valley throne to his current siege of Aspen Angst began in earnest when he encountered Eric Drexler's bipolar vision of nanotechnology, with its manic-depressive alternative futures of utopia and dystopia. Nanotechnology envisages the ultimate creation of new machines and materials, proton by proton, electron by electron, atom by atom. Despite its potential for good, the nightmare is that combined with genetic materials we could create nanobots—self-replicating entities that could multiply themselves into a "gray goo," outperforming photosynthesis and usurping the entire biosphere, including all edible plants and animals.

As terrifying as all this is, Joy's nightmare has one more twist: "The nuclear, biological and chemical technologies used in 20th century weapons of mass destruction were . . . developed in government laboratories," Joy notes. But GNR technologies have "clear commercial uses." They are being developed by "corporate enterprises" which will render them "phenomenally" profitable. Joy has transformed Kurzweil's hopeful vision of freedom and creativity into a Sci-Fi Doomsday scenario.

Joy, whose decades of unfettered research and entrepreneurship have made him what he is today, has fingered the real culprits: capitalism and freedom. Fortunately he has the answer, which he delicately phrases as "relinquishment" of key GNR technologies.

Relinquishment means what it seems to mean: to give up; forgo; abandon not only the use of such technologies but even the basic research that might enable them "to limit the development of the technologies that are too dangerous, by limiting our pursuit of certain types of knowledge." Such relinquishment will require a pre-emptively intrusive, centralized regulatory scheme controlled, of course, by the federal government. The spectacle of one of the world's leading techno-entrepreneurs offering himself as the prime witness for the prosecution (that is, the anti-capitalist left) against his own class is transforming Joy into a celebrity intellectual and political force.

It's amazing what one late night conversation in a bar can set in motion. Joy chose to sound the alarm and call out the cavalry. Perhaps this volume can help get that original debate back on track. In any event, no one should interpret the philosophical criticisms of Kurzweil's views in the following pages as endorsements of Joy's thesis.

Conflicting Visions of the Future—and Reality

Still, Joy's public response does help underscore a profoundly important issue: What will happen when our technological achievements give us Promethean powers—powers once thought the exclusive province of God—just when most of those in charge have ceased to believe in anyone or anything like God?

Many scientists and intellectuals today are very confident that they can do without God. That confidence takes one of two basic forms. The first, cheerful but somewhat hollow, is much in evidence in following pages. Kurzweil, for instance, snatches purpose from a purposeless evolutionary process by *defining* evolution as the purpose of life. The second is ably represented by Joy. For Joy and those of similar persuasion, the future is a source of fear. They can find no answer but brutal exertions of power to terminate the uncertain bounties of human creativity. Their modern naturalistic worldview renders human life and history accidental, unlikely, and sure of a bad end.

Seeing history as a domain of mere chance, they wish to bring it to a halt. Seeing that science cannot prove a negative—guarantee that some invention will not cause a catastrophe—they insist on a "cautionary principle" for new technology that would not have allowed a caveman to build a fire. After all, over the millennia millions of people have died from fire. Seeing that science cannot assure safety, they believe that the endless restlessness and creativity of human beings is a threat rather than an opportunity or a gift.

The human race has prevailed against the plagues and scarcities of its past, not through regulation or "relinquishment" but through creativity and faith. It is chiefly when we give up on freedom and providence, and attempt to calculate and control our destinies through a demiurgic state, that disaster occurs. It is chiefly when we regard the masses as a mob of mouths, accidentally evolved in a random universe, that evil seems inevitable, good incomprehensible, and tyranny indispensable.

To the theist, reality is more than mere chance and mechanistic law. These categories are subsumed by divine freedom and creativity, and become the arena of human freedom and creativity, the proximate sources of technological innovation and wealth. Human creative freedom flourishes in an environment of top-down law and transcendent order, a monotheism that removes the arbitrary from science and denies the ultimate victory of evil in the universe. From such a perspective, one is able to embrace what is good in invention, innovation and technology, while denying them the last word. Without such a viewpoint, one is doomed to lurch between two sides of a false dilemma: Either Promethean anarchy in which we are masters of our own, self-defined but pointless destiny, or servants of a nanny state that must protect us from ourselves and our own teeming ingenuity.

Ray Kurzweil understands and celebrates human freedom and creativity as sources of wealth and fulfillment, and opposes Luddite attempts to stifle them. Forced to choose between Kurzweil and Joy's visions of the future, we would choose Kurzweil's. But we aren't forced to make that choice. Kurzweil and Joy share a naturalistic worldview with many other leading intellectuals. This dramatically restricts their options, and in our opinion doesn't really allow for a resolution of our current dilemma.

In the following chapters, many conclusions follow as logical consequences of implicit naturalistic presuppositions. Since most intellectuals share these assumptions, there's rarely reason to bring them into the light of day. Once stated explicitly, however, it becomes clear how very bright individuals can so heartily disagree.

- Human intelligence is ultimately the product of a process that didn't have us in mind.

So, the only designed—and transcendent—intelligence Kurzweil and others envision is a higher technological intelligence evolving from our own, which itself evolved from an unintelligent process.

- In the final analysis, we must be some material combination of computational software and hardware.

After all, what else could we be?

- When our technology achieves a sufficient level of computational architecture and complexity, it will become conscious, like we are.

Otherwise, human consciousness might be something inexplicable in materialistic categories.

- If we're a carbon-based, complex, computational, collocation of atoms, and we're conscious, then why wouldn't the same be true of a sufficiently complex silicon-based computer?

Given the naturalistic premise, these conclusions seem reasonable. But what if we don't assume the premise?

Accordingly, for Kurzweil the only salvation and the only eschatology are those in which we become one with our own more rapidly evolving, durable and reliable technology. If we seek immortality, we must seek it somewhere downstream from the flow of cosmic evolution, with its ever-accelerating rate of returns. Upstream is only matter in motion.

Kurzweil's seems to be a substitute vision for those who have lost faith in the traditional object of religious belief. It does not despair but rejoices in evolutionary improvement, even if the very notion of improvement remains somewhat alien in a materialistic universe. It's hardly surprising, then, that when he develops this conviction, he eventually appeals to the idea of God. Thus his perspective is closer to human religious intuitions than is the handwringing of Bill Joy or the reductionist Darwinian materialism of the previous century. This makes it intrinsically more interesting and attractive for those, like Kurzweil, who still seek transcendence in an intellectual culture that has lost its faith in the Transcendent. It also makes it worthier of the serious consideration and scrutiny it receives in the following chapters.

1

The Evolution of Mind in the Twenty-First Century

Ray Kurzweil

11.1.04

An Overview of the Next Several Decades

The intelligence of machines—nonbiological entities—will exceed human intelligence early in this next century. By intelligence, I include all the diverse and subtle ways in which humans are intelligent—including musical and artistic aptitude, creativity, physically moving through the world, and even responding to emotion. By 2019, a $1,000 computer will match the processing power of the human brain—about 20 million billion calculations per second. This level of processing power is a necessary but not sufficient condition for achieving human-level intelligence in a machine. Organizing these resources—the "software" of intelligence—will take us to 2029, by which time your average personal computer will be equivalent to a thousand human brains.

Ray Kurzweil is an award-winning inventor, entrepreneur, and author of several books, including the best-seller The Age of Spiritual Machines *(Viking, 1999).*

Once a computer achieves a level of intelligence comparable to human intelligence, it will necessarily soar past it. A key advantage of nonbiological intelligence is that machines can easily share their knowledge. If I learn French, or read War and Peace, I can't readily download that learning to you. You have to acquire that scholarship the same painstaking way that I did. My knowledge, embedded in a vast pattern of neurotransmitter concentrations and interneuronal connections, cannot be quickly accessed or transmitted. But we won't leave out quick downloading ports in our nonbiological equivalents of human neuron clusters. When one computer learns a skill or gains an insight, it can immediately share that wisdom with billions of other machines.

As a contemporary example, we spent years teaching one research computer how to recognize continuous human speech. We exposed it to thousands of hours of recorded speech, corrected its errors, and patiently improved its performance. Finally, it became quite adept at recognizing speech (I dictated most of my recent book to it). Now if you want your own personal computer to recognize speech, it doesn't have to go through the same process; you can just download the fully trained program in seconds. Ultimately, billions of nonbiological entities can be the master of all human and machine acquired knowledge. Computers are also potentially millions of times faster than human neural circuits, and have far more reliable memories.

One approach to designing intelligent computers will be to copy the human brain, so these machines will seem very human. And through nanotechnology, which is the ability to create physical objects atom by atom, they will have humanlike—albeit greatly enhanced—bodies as well. Having human origins, they will claim to be human, and to have human feelings. And being immensely intelligent, they'll be very convincing when they tell us these things. But are these feelings "real," or just *apparently* real? I will discuss this subtle but vital distinction below. First it is important to understand the nature of nonbiological intelligence, and how it will emerge.

Keep in mind that this is not an alien invasion of intelligent machines. It is emerging from within our human-machine civilization.

There will not be a clear distinction between human and machine as we go through the twenty-first century. First of all, we will be putting computers—neural implants—directly into our brains. We've already started down this path. We have ventral posterior nucleus, subthalmic nucleus, and ventral lateral thalamus neural implants to counteract Parkinson's Disease and tremors from other neurological disorders. I have a deaf friend who now can hear what I am saying because of his cochlear implant. Under development is a retina implant that will perform a similar function for blind individuals, basically replacing certain visual processing circuits of the retina and nervous system. Recently scientists from Emory University placed a chip in the brain of a paralyzed stroke victim who can now begin to communicate and control his environment directly from his brain.

In the 2020s, neural implants will not be just for disabled people, and introducing these implants into the brain will not require surgery, but more about that later. There will be ubiquitous use of neural implants to improve our sensory experiences, perception, memory, and logical thinking.

These "noninvasive" implants will also plug us in directly to the World Wide Web. By 2030, "going to a web site" will mean entering a virtual reality environment. The implant will generate the streams of sensory input that would otherwise come from our real senses, thus creating an all-encompassing virtual environment that responds to the behavior of our own virtual body (and those of others) in the virtual environment. This technology will enable us to have virtual reality experiences with other people—or simulated people—without requiring any equipment not already in our heads. And virtual reality will not be the crude experience that one can experience in today's arcade games. Virtual reality will be as realistic, detailed, and subtle as real reality. So instead of just phoning a friend, you can meet in a virtual French café in Paris, or take a walk on a virtual Mediterranean beach, and it will seem very real. People will be able to have any type of experience with anyone—business, social, romantic, sexual—regardless of physical proximity.

The Growth of Computing

To see into the future, we need insight into the past. We need to discern the relevant trends and their interactions. Many projections of the future suffer from three common failures. The first is that people often consider only one or two iterations of advancement in a technology, as if progress would then come to a halt.

A second is focusing on only one aspect of technology, without considering the interactions and synergies from developments in multiple fields (e.g. computational substrates and architectures, software and artificial intelligence, communication, nanotechnology, brain scanning and reverse engineering, amongst others).

By far the most important failure is to fail to adequately take into consideration the accelerating pace of technology. Many predictions do not factor this in with any consistent methodology, if at all. Ten thousand years ago, there was little salient technological change in even a thousand years. A thousand years ago, progress was much faster and a paradigm shift required only a century or two. In the nineteenth century, we saw more technological change than in the nine centuries preceding it. Then in the first twenty years of the twentieth century, we saw more advancement than in all of the nineteenth century. Now, paradigm shifts (and new business models) take place in only a few years time (lately it appears to be even less than that). Just a decade ago, the Internet was in a formative stage and the World Wide Web had yet to emerge

The fact that the successful application of certain innovations (e.g., the laser) may have taken several decades in the past half century does not mean that, going forward, comparable changes will take nearly as long. The type of transformation that required thirty years during the last half century will take only five to seven years going forward. And the pace will continue to accelerate. It is vital to consider the implications of this phenomenon; progress is not linear, but exponential.

This "law of accelerating returns," as I call it, is true of any evolutionary process. It was true of the evolution of life forms, which required billions of years for the first steps (e.g. primitive cells); later on progress accelerated. During the Cambrian explosion, major paradigm shifts took only tens of millions of years. Later on, humanoids developed over a period of only millions of years, and Homo sapiens over a period of hundreds of thousands of years.

With the advent of a technology creating species, the exponential pace became too fast for evolution through DNA-guided protein synthesis and moved on to human-created technology. The first technological steps — sharp edges, fire, the wheel—took tens of thousands of years, and have accelerated ever since.

Technology is evolution by other means. It is the cutting edge of evolution today, moving far faster than DNA-based evolution. However, unlike biological evolution, technology is not a "blind watchmaker." (Actually, I would prefer the phrase "mindless watchmaker" as more descriptive and less insensitive to the visually impaired.) The process of humans creating technology, then, is a "mindful watchmaker." An implication is that we do have the ability (and the responsibility) to guide this evolutionary process in a constructive direction.

Technology goes beyond mere toolmaking; it is a process of creating ever more powerful technology using the tools from the previous round of innovation. In this way, human technology is distinguished from the toolmaking of other species. There is a record of each stage of technology, and each new stage of technology builds on the order of the previous stage. Technology, therefore, is a continuation of the evolutionary process that gave rise to the technology creating species in the first place.

It is critical when considering the future to use a systematic methodology that considers these three issues: (i) iterations of technological progress do not just stop at an arbitrary point, (ii) diverse developments interact, and, most importantly, (iii) the pace of technological innovation accelerates. This third item can be quantified, which I discuss in the sidebar below. Although these formulas are not perfect models, they do provide a framework for considering future developments. I've used this methodology for the past twenty

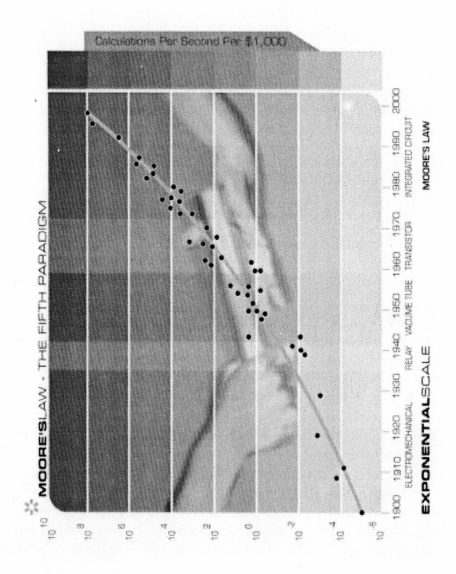

years, and the predictions derived from this method in the 1980s have held up rather well.

One very important trend is referred to as "Moore's Law." Gordon Moore, one of the inventors of integrated circuits, and then chairman of Intel, noted in the mid-1970s that we could squeeze twice as many transistors on an integrated circuit every twenty-four months. The implication is that computers, which are built from integrated circuits, are doubling in power every two years. Lately, the rate has been even faster.

After sixty years of devoted service, Moore's Law will die a dignified death no later than the year 2019. By that time, transistor features will be just a few atoms in width, and the strategy of ever finer photolithography will have run its course. So, will that be the end of the exponential growth of computing?

Don't bet on it.

If we plot the speed (in instructions per second) per $1000 (in constant dollars) of 49 famous calculators and computers spanning the entire twentieth century, we note some interesting observations.

It is important to note that Moore's Law of Integrated Circuits was not the first, but the fifth paradigm to provide accelerating price-performance. Computing devices have been consistently multiplying in power (per unit of time) from the mechanical calculating devices used in the 1890 U.S. census, to Turing's relay-based "Robinson" machine that cracked the Nazi enigma code, to the CBS vacuum tube computer that predicted the election of Eisenhower, to the transistor-based machines used in the first space launches, to the integrated-circuit-based personal computer which I used to dictate (and automatically transcribe) this chapter.

But I noticed something else surprising. When I plotted the 49 machines on a logarithmic graph (where a straight line means exponential growth), I didn't get a straight line. What I got was another exponential curve. In other words, there's exponential growth in the rate of exponential growth. Computer speed (per unit cost) doubled every three years between 1910 and 1950, doubled every two years between 1950 and 1966, and is now doubling every year.

Wherefrom Moore's Law:
The Law of Accelerating Returns

Where does Moore's Law come from? What is behind this remarkably predictable phenomenon? I have seen relatively little written about the ultimate source of this trend. Is it just "a set of industry expectations and goals," as Randy Isaac, head of basic science at IBM, contends?

In my view, it is one manifestation (among many) of the exponential growth of the evolutionary process that is technology. Just as the pace of an evolutionary process accelerates, the "returns" (i.e., the output, the products) of an evolutionary process grow exponentially. The exponential growth of computing is a marvelous quantitative example of the exponentially growing returns from an evolutionary process. We can also express the exponential growth of computing in terms of an accelerating pace: It took 90 years to achieve the first Multiple in Power (MIP) per thousand dollars; now we add a MIP per thousand dollars every day.

Moore's Law narrowly refers to the number of transistors on an integrated circuit of fixed size, and sometimes has been expressed even more narrowly in terms of transistor feature size. But rather than feature size (which is only one contributing factor), or even number of transistors, I think the most salient measure to track is computational speed per unit cost. This takes into account many levels of "cleverness" (i.e., innovation, which is to say technological evolution). In addition to all of the innovation in integrated circuits, there are multiple layers of innovation in computer design, e.g., pipelining, parallel processing, instruction look-ahead, instruction and memory caching, etc.

From the above chart, we see that the exponential growth of computing didn't start with integrated circuits (around 1958), or even transistors (around 1947), but goes back to the electromechanical calculators used in the 1890 and 1900 U.S. census. This chart spans at least five distinct paradigms of computing (electromechanical calculators, relay-based computers, vacuum tube-based computers, discrete transistor-based computers, and finally microprocessors), of which Moore's Law pertains to only the latest one.

It's obvious what the sixth paradigm will be after Moore's Law runs out of steam before 2019 (because before then transistor feature sizes will be just a few atoms in width). Chips today are flat (although it does require up to 20 layers of material to produce one layer of circuitry). Our brain, in contrast, is organized in three dimensions. We live in a three-dimensional world; why not use the third dimension? There are many technologies in the wings that build circuitry in three dimensions. Nanotubes, for example, which are already working in laboratories, build circuits from pentagonal arrays of carbon atoms. One cubic inch of nanotube circuitry would be a million times more powerful than the human brain.

Thus the (double) exponential growth of computing is broader than Moore's Law. And this accelerating growth of computing is, in turn, part of a yet broader phenomenon discussed above, the accelerating pace of any evolutionary process. In my book, I discuss the link between the pace of a process and the degree of chaos versus order in the process. For example, in cosmological history, the Universe started with little chaos, so the first three major paradigm shifts (the emergence of gravity, the emergence of matter, and the emergence of the four fundamental forces) all occurred in the first billionth of a second; now with vast chaos, cosmological paradigm shifts take billions of years.

Observers are quick to criticize extrapolations of an exponential trend on the basis that the trend is bound to run out of "resources." The classical example is when a species happens upon a new habitat (e.g., rabbits in Australia), the species' numbers will grow exponentially for a time, but then hit a limit when resources such as food and space run out. But the resources underlying the exponential growth of an evolutionary process are relatively unbounded: (i) the (ever growing) order of the evolutionary process itself, and (ii) the chaos of the environment in which the evolutionary process takes place and which provides the options for further diversity.

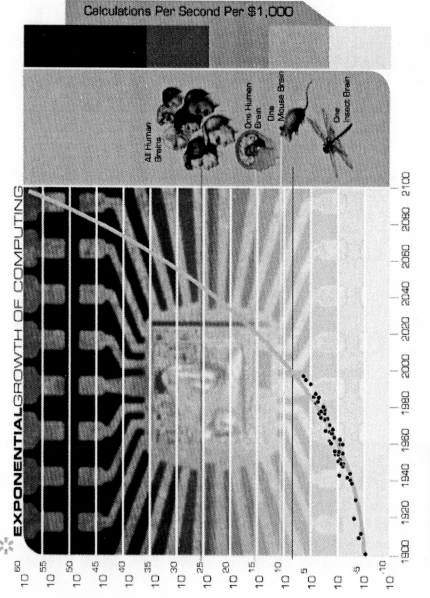

We also need to distinguish between the S curve (very slow virtually unnoticeable growth followed by very rapid growth followed by the growth leveling off and reaching an asymptote) that is characteristic of any specific technological paradigm and the continuing exponential growth that is characteristic of the ongoing evolutionary process of technology. Specific paradigms, such as Moore's Law (i.e., achieving faster and denser computation through shrinking transistor sizes on an integrated circuit), do ultimately reach levels at which exponential growth is no longer feasible. Thus Moore's Law is an S curve. But the growth of computation is an ongoing exponential. What turns the S curve (of any specific paradigm) into a continuing exponential is paradigm shift (also called innovation), in which a new paradigm (e.g., three-dimensional circuits) takes over when the old paradigm approaches its natural limit. This has already happened at least four times in the history of computation. This difference also distinguishes the toolmaking of non-human species, in which the mastery of a toolmaking (or using) skill by each animal is characterized by an S-shaped learning curve, and human-created technology, which has been following an exponential pattern of growth and acceleration since its inception.

I discuss all of this in more detail in the first couple of chapters of my book.

In the sidebar below, I include a mathematical model of the law of accelerating returns as it pertains to the exponential growth of computing. The formulas below result in the following graph of the continued growth of computation. This graph matches the available data for the twentieth century and provides projections for the twenty-first century. Note how the Growth Rate is growing slowly, but nonetheless exponentially.

Another technology trend that will have important implications for the twenty-first century is miniaturization. A related analysis can be made of this trend which shows that the salient implementation sizes of a broad range of technology are shrinking, also at a double exponential rate. At present, we are shrinking technology by a factor of approximately 5.6 per linear dimension per decade.

THE LAW OF ACCELERATING RETURNS APPLIED TO THE GROWTH OF COMPUTATION

The following provides a brief overview of the law of accelerating returns as it applies to the double exponential growth of computation. This model considers the impact of the growing power of the technology to foster its own next generation. For example, with more powerful computers and related technology, we have the tools and the knowledge to design yet more powerful computers, and to do so more quickly.

Note that the data for the year 2000 and beyond assume neural net connection calculations as it is expected that this type of calculation will dominate, particularly in emulating human brain functions. This type of calculation is less expensive than conventional (e.g., Pentium III) calculations by a factor of 10 (particularly if implemented using digital controlled analog electronics, which would correspond well to the brain's digital controlled analog electrochemical processes). A factor of 10 translates into approximately 3 years (today) and less than 3 years later in the twenty-first century.

My estimate of brain capacity is 100 billion neurons times an average 1,000 connections per neuron (with the calculations taking place primarily in the connections) times 200 calculations per second. Although these estimates are conservatively high, one can find higher and lower estimates. However, even much higher (or lower) estimates by orders of magnitude only shift the prediction by a relatively small number of years.

Some salient dates from this analysis include the following:

We achieve one Human Brain capability ($2 * 10^{16}$ cps) for $1,000 around the year 2023.

We achieve one Human Brain capability ($2 * 10^{16}$ cps) for one cent around the year 2037.

We achieve one Human Race capability ($2 * 10^{26}$ cps) for $1,000 around the year 2049.

We achieve one Human Race capability ($2 * 10^{26}$ cps) for one cent around the year 2059.

The Model considers the following variables:

V: Velocity (i.e., power) of computing (measured in CPS/unit cost)

W: World Knowledge as it pertains to designing and building computational devices

t: Time

The assumptions of the model are:

(1) $V = C1 * W$

In other words, computer power is a linear function of the knowledge of how to build computers. This is actually a conservative assumption. In general, innovations improve V (computer power) by a multiple, not in an additive way. Independent innovations multiply each other's effect. For example, a circuit advance such as CMOS, a more efficient IC wiring methodology, and a processor innovation such as pipelining all increase V by independent multiples.

(2) W = C2 * Integral (0 to t) V

In other words, W (knowledge) is cumulative, and the instantaneous increment to knowledge is proportional to V.

This gives us:

W = C1 * C2 * Integral (0 to t) W

W = C1 * C2 * C3 ^ (C4 * t)

V = C1 ^ 2 * C2 * C3 ^ (C4 * t)

(Note on notation: a^b means a raised to the b power.)

Simplifying the constants, we get:

V = Ca * Cb ^ (Cc * t)

So this is a formula for "accelerating" (i.e., exponentially growing) returns, a "regular Moore's Law."

As I mentioned above, the data shows exponential growth in the rate of exponential growth. (We doubled computer power every three years early in the twentieth century, every two years in the middle of the century, and close to every one year during the 1990s.)

Let's factor in another exponential phenomenon, which is the growing resources for computation. Not only is each (constant cost) device getting more powerful as a function of W, but the resources deployed for computation are also growing exponentially.

We now have:

N: Expenditures for computation

V = C1 * W (as before)

N = C4 ^ (C5 * t) (Expenditure for computation is growing at its own exponential rate)

W = C2 * Integral (0 to t) (N * V)

As before, world knowledge is accumulating, and the instantaneous increment is proportional to the amount of computation, which equals the resources deployed for computation (N) * the power of each (constant cost) device.

This gives us:

W = C1 * C2 * Integral(0 to t) (C4 ^ (C5 * t) * W)

W = C1 * C2 * (C3 ^ (C6 * t)) ^ (C7 * t)

V = C1 ^ 2 * C2 * (C3 ^ (C6 * t)) ^ (C7 * t)

Simplifying the constants, we get:

V = Ca * (Cb ^ (Cc * t)) ^ (Cd * t)

This is a double exponential—an exponential curve in which the rate of exponential growth is growing at a different exponential rate.

Now let's consider real-world data. Considering the data for actual calculating devices and computers during the twentieth century:

CPS/$1K: Calculations Per Second for $1,000

Twentieth century computing data matches:

CPS/$1K = $10^{\wedge}(6.00*((20.40/6.00)^{\wedge}((A13-1900)/100))-11.00)$

We can determine the growth rate over a period of time:

Growth Rate =$10^{\wedge}((LOG(CPS/\$1K$ for Current Year)—LOG(CPS/$1K$ for Previous Year))/(Current Year—Previous Year))

Human Brain = 100 Billion ($10^{\wedge}11$) neurons * 1000 ($10^{\wedge}3$) Connections/Neuron * 200 (2 * $10^{\wedge}2$) Calculations Per Second Per Connection = 2 * $10^{\wedge}16$ Calculations Per Second

Human Race = 10 Billion ($10^{\wedge}10$) Human Brains = 2 * $10^{\wedge}26$ Calculations Per Second

These formulas produce the graph above.

In a process, the time interval between salient events expands or contracts along with the amount of chaos. This relationship is one key to understanding the reason that the exponential growth of computing will survive the demise of Moore's Law. Evolution started with vast chaos and little effective order, so early progress was slow. But evolution creates ever-increasing order. That is, after all, the essence of evolution. Order is the opposite of chaos, so when order in a process increases—as is the case for evolution—time speeds up. I call this important sub-law the "law of accelerating returns," to contrast it with a better known law in which returns diminish.

Computation represents the essence of order in technology. Being subject to the evolutionary process that is technology, it too grows exponentially. There are many examples of the exponential growth in technological speeds and capacities. For example, when the human genome scan started twelve years ago, genetic sequencing speeds were so slow that without speed increases the project would have required thousands of years, yet it is now completing on schedule in under fifteen years. Other examples include the accelerating price-performance of all forms of computer memory, the exponential growing bandwidth of communication technologies (electronic, optical, as well as wireless), the rapidly increasing speed and resolution of human brain scanning, the miniaturization of technology, and many others. If we view the exponential growth of computation in its proper perspective, as one example of many of the law of accelerating returns, then we can confidently predict its continuation.

A sixth paradigm will take over from Moore's Law, just as Moore's Law took over from discrete transistors, and vacuum tubes before that. There are many emerging technologies for new computational substrates. In addition to nanotubes, several forms of computing at the molecular level are working in laboratories. There are more than enough new computing technologies now being researched, including three-dimensional chips, optical computing, crystalline computing, DNA computing, and quantum computing, to keep the law of accelerating returns going for a long time.

So where will this take us?

IBM's "Blue Gene" supercomputer, scheduled to be completed by 2005, is projected to provide 1 million billion calculations per second, already one-twentieth of the capacity of the human brain. By the year 2019, your $1,000 personal computer will have the processing power of the human brain—20 million billion calculations per second (100 billion neurons times 1,000 connections per neuron times 200 calculations per second per connection). By 2029, it will take a village of human brains (about a thousand) to match $1,000 of computing. By 2050, $1,000 of computing will equal the processing power of all human brains on Earth. Of course, this only includes those brains still using carbon-based neurons. While human neurons are wondrous creations in a way, we wouldn't design computing circuits the same way. Our electronic circuits are already more than 10 million times faster than a neuron's electrochemical processes. Most of the complexity of a human neuron is devoted to maintaining its life support functions, not its information processing capabilities. Ultimately, we will need to port our mental processes to a more suitable computational substrate. Then our minds won't have to stay so small, being constrained as they are today to a mere hundred trillion neural connections each operating at a ponderous 200 digitally controlled analog calculations per second.

A careful consideration of the law of time and chaos, and its key sublaw, the law of accelerating returns, shows that the exponential growth of computing is not like those other exponential trends that run out of resources. The two resources it needs—the growing order of the evolving technology itself, and the chaos from which an evolutionary process draws its options for further diversity—are without practical limits, at least not limits that we will encounter in the next century.

The Intuitive Linear View versus The Historical Exponential View

Many long range forecasts of technical feasibility in future time periods dramatically underestimate the power of future technology because they are based on what I call the "intuitive linear" view of technological progress rather than the "historical exponential view."

To express this another way, it is not the case that we will experience a hundred years of progress in the twenty-first century; rather we will witness on the order of twenty thousand years of progress (from the linear perspective, that is).

When people think of a future period, they intuitively assume that the current rate of progress will continue for the period being considered. However, careful consideration of the pace of technology shows that the rate of progress is not constant, but it is human nature to adapt to the changing pace, so the intuitive view is that the pace will continue at the current rate. Even for those of us who have lived through a sufficiently long period of technological progress to experience how the pace increases over time, our unexamined intuition nonetheless provides the impression that progress changes at the rate that we have experienced recently. A salient reason for this is that an exponential curve approximates a straight line when viewed for a brief duration. So even though the rate of progress in the very recent past (e.g., this past year) is far greater than it was ten years ago (let alone a hundred or a thousand years ago), our memories are nonetheless dominated by our very recent experience. Since the rate has not changed significantly in the very recent past (because a very small piece of an exponential curve is approximately straight), it is an understandable misperception to view the pace of change as a constant. It is typical, therefore, that even sophisticated commentators, when considering the future, extrapolate the current pace of change over the next ten years or hundred years to determine their expectations. This is why I call this way of looking at the future the "intuitive linear" view.

But any serious consideration of the history of technology shows that technological change is at least exponential, not linear. There are a great many examples of this which I have discussed above. One can examine this data in many different ways, and on many different time scales, and for a wide variety of different phenomena, and the (at least) double exponential growth implied by the law of accelerating returns applies. The law of accelerating returns does not rely on an assumption of the continuation of Moore's law, but is based on a rich model of diverse technological processes. What it clearly shows is that technology, particularly the pace of technological change, advances (at least) exponentially, not linearly, and has been doing so since the advent of technology, indeed since the advent of evolution on Earth.

Most technology forecasts ignore altogether this "historical exponential view" of technological progress and assume instead the "intuitive linear view." Although the evidence is compelling, it still requires study and modeling of many diverse events to see this exponential aspect. That is why people tend to overestimate what can be achieved in the short term (because we tend to leave out necessary details), but underestimate what can be achieved in the long term (because the exponential growth is ignored).

This observation also applies to paradigm shift rates, which are currently doubling (approximately) every decade; that is paradigm shift times are halving every decade (and this rate is also changing slowly, but nonetheless exponentially). So, the technological progress in the twenty-first century will be equivalent to what would require (in the linear view) on the order of twenty thousand years. In terms of the growth of computing, the comparison is even more dramatic.

The Software of Intelligence

So far, I've been talking about the hardware of computing. The software is even more salient. Achieving the computational capacity of the human brain, or even villages and nations of human brains will not automatically produce human levels of capability. It is a necessary but not sufficient condition. The organization and content of these resources—the *software* of intelligence—is also critical.

There are a number of compelling scenarios to capture higher levels of intelligence in our computers, and ultimately human levels and beyond. We will be able to evolve and train a system combining massively parallel neural nets with other paradigms to understand language and model knowledge, including the ability to read and model the knowledge contained in written documents. Unlike many

contemporary "neural net" machines, which use mathematically simplified models of human neurons, more advanced neural nets are
already using highly detailed models of human neurons, including
detailed nonlinear analog activation functions and other salient details. Although the ability of today's computers to extract and learn
knowledge from natural language documents is limited, their capabilities in this domain are improving rapidly. Computers will be able
to read on their own, understanding and modeling what they have
read, by the second decade of the twenty-first century. We can then
have our computers read all of the world's literature—books, maga-
zines, scientific journals, and other available material. Ultimately,
the machines will gather knowledge on their own by venturing into
the physical world, drawing from the full spectrum of media and
information services, and sharing knowledge with each other (which
machines can do far more easily than their human creators).

Once a computer achieves a human level of intelligence, it will
necessarily soar past it. Since their inception, computers have significantly exceeded human mental dexterity in their ability to remember and process information. A computer can remember billions or
even trillions of facts perfectly, while we are hard pressed to remember a handful of phone numbers. A computer can quickly search a
data base with billions of records in fractions of a second. As I mentioned earlier, computers can readily share their knowledge. The combination of human level intelligence in a machine with a computer's
inherent superiority in the speed, accuracy and sharing ability of its
memory will be formidable.

Reverse Engineering the Human Brain

The most compelling scenario for mastering the software of intelligence is to tap into the blueprint of the best example we can get our
hands on of an intelligent process. There is no reason why we cannot
reverse engineer the human brain, and essentially copy its design. It
took its original designer several billion years to develop. And it's
not even copyrighted.

The most immediately accessible way to accomplish this is through destructive scanning: we take a frozen brain, preferably one frozen just slightly before rather than slightly after it was going to die anyway, and examine one brain layer—one very thin slice—at a time. We can readily see every neuron and every connection and every neurotransmitter concentration represented in each synapse-thin layer.

Human brain scanning has already started. A condemned killer allowed his brain and body to be scanned and you can access all 10 billion bytes of him on the Internet. He has a 25 billion byte female companion on the site as well in case he gets lonely. This scan is not high enough resolution for our purposes, but then we probably don't want to base our templates of machine intelligence on the brain of a convicted killer, anyway.

But scanning a frozen brain is feasible today, albeit not yet at a sufficient speed or bandwidth, but again, the law of accelerating returns will provide the requisite speed of scanning, just as it did for the human genome scan.

We also have noninvasive scanning techniques today, including high-resolution magnetic resonance imaging (MRI) scans, optical imaging, near-infrared scanning, and other noninvasive scanning technologies, that are capable in certain instances of resolving individual somas, or neuron cell bodies. Brain scanning technologies are increasing their resolution with each new generation, just what we would expect from the law of accelerating returns. Future generations will enable us to resolve the connections between neurons, and to peer inside the synapses and record the neurotransmitter concentrations.

We can peer inside someone's brain today with noninvasive scanners, which are increasing their resolution with each new generation of this technology. There are a number of technical challenges in accomplishing this, including achieving suitable resolution, bandwidth, lack of vibration, and safety. For a variety of reasons it is easier to scan the brain of someone recently deceased than of someone still living. It is easier to get someone deceased to sit still, for one thing. But noninvasively scanning a living brain will ultimately become feasible as MRI, optical, and other scanning technologies continue to improve in resolution and speed.

In fact, the driving force behind the rapidly improving capability of noninvasive scanning technologies is again the law of accelerating returns, because it requires massive computational ability to build the high-resolution three-dimensional images. The exponentially increasing computational ability provided by the law of accelerating returns (and for another 10 to 20 years, Moore's Law) will enable us to continue to rapidly improve the resolution and speed of these scanning technologies.

Scanning from Inside

To capture every salient neural detail of the human brain, the most practical approach will be to scan it from inside. By 2030, "nanobot" (i.e., nano-robot) technology will be viable, and brain scanning will be a prominent application. Nanobots are robots that are the size of human blood cells, or even smaller. Billions of them could travel through every brain capillary and scan every salient neural detail from up close. Using high-speed wireless communication, the nanobots would communicate with each other, and with other computers that are compiling the brain scan database (in other words, the nanobots will all be on a wireless local area network).

This scenario involves only capabilities we can touch and feel today. We already have technology capable of producing very high-resolution scans provided that the scanner is physically proximate to the neural features. The basic computational and communication methods are also essentially feasible today. The primary features that are not yet practical are nanobot size and cost. As I discussed above, we can project the exponentially declining cost of computation. Miniaturization is another readily predictable aspect of the law of accelerating returns. Already being developed at the University of California at Berkeley are tiny flying robots called "smart dust," which are approximately one millimeter wide (about the size of a grain of sand), and capable of flying, sensing, computing, and communicating using tiny lasers and hinged micro-flaps and micro-mirrors. The size of electronics and robotics will continue to shrink at an exponential rate, currently by a factor of 5.6 per linear dimension per decade. We can conservatively expect, therefore, the requisite nanobot technology by around 2030. Because of its ability to place each scanner in very close physical proximity to every neural feature, nanobot-based scanning will be more practical than scanning the brain from outside.

How to Use Your Brain Scan

How will we apply the thousands of trillions of bytes of information derived from each brain scan? One approach is to use the results to design more intelligent parallel algorithms for our machines, particularly those based on one of the neural net paradigms. With this approach, we don't have to copy every single connection. There is a great deal of repetition and redundancy within any particular brain region. Although the information contained in a human brain would require thousands of trillions of bytes of information (on the order of 100 billion neurons times an average of 1,000 connections per neuron, each with multiple neurotransmitter concentrations and connection data), the design of the brain is characterized by a human genome of only about a billion bytes.

Furthermore, most of the genome is redundant, so the initial design of the brain is characterized by approximately one hundred million bytes, about the size of Microsoft Word. Of course, the complexity of our brains greatly increases as we interact with the world (by a factor of more than ten million). It is not necessary, however, to capture each detail in order to reverse engineer the salient digital-analog algorithms. With this information, we can design simulated nets that operate similarly. There are already multiple efforts under way to scan the human brain and apply the insights derived to the design of intelligent machines. The ATR (Advanced Telecommunications Research) Lab in Kyoto, Japan, for example, is building a silicon brain with 1 billion neurons. Although this is 1% of the number of neurons in a human brain, the ATR neurons operate at much faster speeds.

After the algorithms of a region are understood, they can be refined and extended before being implemented in synthetic neural equivalents. For one thing, they can be run on a computational substrate that is already more than ten million times faster than neural circuitry. And we can also throw in the methods for building intelligent machines that we already understand.

Downloading the Human Brain

Perhaps a more interesting approach than this scanning-the-brain-to-understand-it scenario is *scanning-the-brain-to-download-it*. Here we scan someone's brain to map the locations, interconnections, and contents of all the somas, axons, dendrites, presynaptic vesicles, neurotransmitter concentrations, and other neural components and levels. Its entire organization can then be re-created on a neural computer of sufficient capacity, including the contents of its memory.

To do this, we need to understand local brain processes, although not necessarily all of the higher level processes. Scanning a brain with sufficient detail to download it may sound daunting, but so did the human genome scan. All of the basic technologies exist today, just not with the requisite speed, cost, and size, but these are the attributes that are improving at a double exponential pace.

The computationally salient aspects of individual neurons are complicated, but definitely not beyond our ability to accurately model. For example, Ted Berger and his colleagues at Hedco Neurosciences have built integrated circuits that precisely match the digital and analog information processing characteristics of neurons, including clusters with hundreds of neurons. Carver Mead and his colleagues at CalTech have built a variety of integrated circuits that emulate the digital-analog characteristics of mammalian neural circuits.

A recent experiment at San Diego's Institute for Nonlinear Science demonstrates the potential for electronic neurons to precisely emulate biological ones. Neurons (biological or otherwise) are a prime example of what is often called "chaotic computing." Each neuron acts in an essentially unpredictable fashion. When an entire network of neurons receives input (from the outside world or from other networks of neurons), the signaling amongst them appears at first to be frenzied and random. Over time, typically a fraction of a second or so, the chaotic interplay of the neurons dies down, and a stable pattern emerges. This pattern represents the "decision" of the neural network. If the neural network is performing a pattern recognition task (which, incidentally, comprises more than 95% of the activity in the human brain), then the emergent pattern represents the appropriate recognition.

So the question addressed by the San Diego researchers was whether electronic neurons could engage in this chaotic dance alongside biological ones. They hooked up their artificial neurons with those from spiney lobsters in a single network, and their hybrid biological-nonbiological network performed in the same way (i.e., chaotic interplay followed by a stable emergent pattern) and with the same type of results as an all biological net of neurons. Essentially, the biological neurons accepted their electronic peers. It indicates that their mathematical model of these neurons was reasonably accurate.

There are many projects around the world, which are creating nonbiological devices and which recreate in great detail the functionality of human neuron clusters, and the accuracy and scale of these neuron clusters replications are rapidly increasing. We started with functionally equivalent recreations of single neurons, then clusters of tens, then hundreds, and now thousands. Scaling up technical processes at an exponential pace is what technology is good at.

As the computational power to emulate the human brain becomes available—we're not there yet, but we will be there within a couple of decades—projects already under way to scan the human brain will be accelerated, with a view both to understand the human brain in general, as well as providing a detailed description of the contents and design of specific brains. By the third decade of the twenty-first century, we will be in a position to create highly detailed and complete maps of all relevant features of all neurons, neural connections and synapses in the human brain, all of the neural details that play a role in the behavior and functionality of the brain, and to recreate these designs in suitably advanced neural computers.

Is the Human Brain Different from a Computer?

The answer depends on what we mean by the word "computer." Certainly the brain uses very different methods from conventional contemporary computers. Most computers today are all digital and perform one (or perhaps a few) computation(s) at a time at extremely high speed. In contrast, the human brain combines digital and analog methods with most computations performed in the analog domain. The brain is massively parallel, performing on the order of a hundred trillion computations at the same time, but at extremely slow speeds.

With regard to digital versus analog computing, we know that digital computing can be functionally equivalent to analog computing (although the reverse is not true), so we can perform all of the capabilities of a hybrid digital—analog network with an all digital computer. On the other hand, there is an engineering advantage to analog circuits in that analog computing is potentially thousands of times more efficient. An analog computation can be performed by a few transistors, or, in the case of mammalian neurons, specific electrochemical processes. A digital computation in contrast requires thousands or tens of thousands of transistors. So there is a significant efficiency advantage to emulating the brain's analog methods.

The massive parallelism of the human brain is the key to its pattern recognition abilities, which reflects the strength of human thinking. As I discussed above, mammalian neurons engage in a chaotic dance, and if the neural network has learned its lessons well, then a stable pattern will emerge reflecting the network's decision. There is no reason why our nonbiological functionally-equivalent recreations of biological neural networks cannot be built using these same principles, and indeed there are dozens of projects around the world that have succeeded in doing this. My own technical field is pattern recognition, and the projects that I have been involved in for over thirty years use this form of chaotic computing. Particularly successful examples are Carver Mead's neural chips, which are highly parallel, use digital controlled analog computing, and are intended as functionally similar recreations of biological networks.

As we create nonbiological but functionally equivalent recreations of biological neural networks ranging from clusters of dozens of neurons up to entire human brains and beyond, we can combine the qualities of human thinking with certain advantages of machine intelligence. My human knowledge and skills exist in my brain as vast patterns of interneuronal connections, neurotransmitter concentrations, and other neural elements. As I mentioned at the beginning of this chapter, there are no quick downloading ports for these patterns in our biological neural networks, but as we build nonbiological equivalents, we will not leave out the ability to quickly load patterns representing knowledge and skills.

Although it is remarkable that as complex and capable an entity as the human brain evolved through natural selection, aspects of its design are nonetheless extremely inefficient. Neurons are very bulky devices and at least ten million times slower in their information processing than electronic circuits. As we combine the brain's pattern recognition methods derived from high-resolution brain scans and reverse engineering efforts with the knowledge sharing, speed, and memory accuracy advantages of nonbiological intelligence, the combination will be formidable.

Objective and Subjective

Although I anticipate that the most common application of the knowledge gained from reverse engineering the human brain will be creating more intelligent machines that are not necessarily modeled on specific individuals, the scenario of scanning and reinstantiating all of the neural details of a particular person raises the most immediate questions of identity. Let's consider the question of what we will find when we do this.

We have to consider this question on both the objective and subjective levels. "Objective" means everyone except me, so let's start with that. Objectively, when we scan someone's brain and reinstantiate their personal mind file into a suitable computing medium, the newly emergent "person" will appear to other observers to have very much the same personality, history, and memory as the person originally scanned. That is, once the technology has been refined and perfected. Like any new technology, it won't be perfect at first. But ultimately, the scans and recreations will be very accurate and realistic.

Interacting with the newly instantiated person will feel like interacting with the original person. The new person will claim to be that same old person and will have a memory of having been that person. The new person will have all of the patterns of knowledge, skill, and personality of the original. We are already creating functionally equivalent recreations of neurons and neuron clusters with sufficient accuracy that biological neurons accept their nonbiological equivalents and work with them as if they were biological. There are no natural limits that prevent us from doing the same with the hundred billion neuron cluster we call the human brain. *Claiming?*

Subjectively, the issue is more subtle and profound, but first we need to reflect on one additional objective issue: our physical self.

The Importance of Having a Body

Consider how many of our thoughts and thinking is directed towards our body and its survival, security, nutrition, image, not to mention affection, sexuality, and reproduction. Many, if not most, of the goals we attempt to advance using our brains have to do with our bodies: protecting them, providing them with fuel, making them attractive, making them feel good, providing for their myriad needs and desires. Some philosophers maintain that achieving human level intelligence is impossible without a body. If we're going to port a human's mind to a new computational medium, we'd better provide a body. A disembodied mind will quickly get depressed.

There are a variety of bodies that we will provide for our machines, and that they will provide for themselves: bodies built through nanotechnology (an emerging field devoted to building highly complex physical entities atom by atom), virtual bodies (that exist only in virtual reality), bodies comprised of swarms of nanobots.

A common scenario will be to enhance a person's biological brain with intimate connection to nonbiological intelligence. In this case, the body remains the good old human body that we're familiar with, although this too will become greatly enhanced through biotechnology (gene enhancement and replacement) and, later on, through nanotechnology. A detailed examination of twenty-first century bodies is beyond the scope of this chapter, but is examined in chapter seven of my recent book *The Age of Spiritual Machines.*

So Just Who are These People?

To return to the issue of subjectivity, consider: Is the reinstantiated mind the same consciousness as the person we just scanned? Are these "people" conscious at all? Is this a mind or just a brain?

Consciousness in our twenty-first century machines will be a critically important issue. But it is not easily resolved, or even readily understood. People tend to have strong views on the subject, and often just can't understand how anyone else could possibly see the issue from a different perspective. Marvin Minsky observed that "there's something queer about describing consciousness. Whatever people mean to say, they just can't seem to make it clear."

We don't worry, at least not yet, about causing pain and suffering to our computer programs. But at what point do we consider an entity, a process, to be conscious, to feel pain and discomfort, to have its own intentionality, its own free will? How do we determine if an entity is conscious; if it has subjective experience? How do we distinguish a process that is conscious from one that just acts *as if* it is conscious?

We can't simply ask it. If it says, "Hey I'm conscious," does that settle the issue? No, we have computer games today that effectively do that, and they're not terribly convincing.

How about if the entity *is* very convincing and compelling when it says, "I'm lonely, please keep me company"? Does that settle the issue?

If we look inside its circuits, and see essentially the identical kinds of feedback loops and other mechanisms in its brain that we see in a human brain (albeit implemented using nonbiological equivalents), does that settle the issue?

And just who are these people in the machine, anyway? The answer will depend on who you ask. If you ask the people in the machine, they will strenuously claim to be the original persons. For example, if we scan—let's say myself—and record the exact state, level, and position of every neurotransmitter, synapse, neural connection, and every other relevant detail, and then reinstantiate this massive data base of information (which I estimate at thousands of trillions of bytes) into a neural computer of sufficient capacity, the person that then emerges in the machine will think that he is (and had been) me. He will say "I grew up in Queens, New York, went to college at MIT, stayed in the Boston area, sold a few artificial intelligence companies, walked into a scanner there, and woke up in the machine here. Hey, this technology really works."

But wait. Is this really me? For one thing, old biological Ray (that's me) still exists. I'll still be here in my carbon-cell-based brain. Alas, I will have to sit back and watch the new Ray succeed in endeavors that I could only dream of.

A Thought Experiment

Let's consider the issue of just who I am, and who the new Ray is a little more carefully. First of all, am I the stuff in my brain and body?

Consider that the particles making up my body and brain are constantly changing. We are not at all permanent collections of particles. The cells in our bodies turn over at different rates, but the particles (e.g., atoms and molecules) that comprise our cells are exchanged at a very rapid rate. I am just not the same collection of particles that I was even a month ago. It is the patterns of matter and energy that are semipermanent (that is, changing only gradually), but our actual material content is changing constantly, and very quickly. We are rather like the patterns that water makes in a stream. The rushing water around a formation of rocks makes a particular, unique pattern. This pattern may remain relatively unchanged for hours, even years. Of course, the actual material constituting the pattern—the water—is replaced in milliseconds. The same is true for Ray Kurzweil. Like the water in a stream, my particles are constantly changing, but the pattern that people recognize as Ray has a reasonable level of continuity. This argues that we should not associate our fundamental identity with a specific set of particles, but rather the pattern of matter and energy that we represent. Many contemporary philosophers seem partial to this "identify from pattern" argument.

But wait. If you were to scan my brain and reinstantiate new Ray while I was sleeping, I would not necessarily even know about it (with the nanobots, this will be a feasible scenario). If you then come to me, and say, "Good news, Ray, we've successfully reinstantiated your mind file, so we won't be needing your old brain anymore," I may suddenly realize the flaw in the "identity from pattern" argument. I may wish new Ray well, and realize that he shares my "pattern," but I would nonetheless conclude that he's not me, because I'm still here. How could he be me? After all, I would not necessarily know that he even existed.

Let's consider another perplexing scenario. Suppose I replace a small number of biological neurons with functionally equivalent nonbiological ones (they may provide certain benefits such as greater reliability and longevity, but that's not relevant to this thought experiment). After I have this procedure performed, am I still the same person? My friends certainly think so. I still have the same self-deprecating humor, the same silly grin—yes, I'm still the same guy.

It should be clear where I'm going with this. Bit by bit, region by region, I ultimately replace my entire brain with essentially identical (perhaps improved) nonbiological equivalents (preserving all of the neurotransmitter concentrations and other details that represent my learning, skills, and memories). At each point, I feel the procedures were successful. At each point, I feel that I am same guy. After each procedure, I claim to be the same guy. My friends concur. There is no old Ray and new Ray, just one Ray, one that never appears to fundamentally change.

But consider this. This gradual replacement of my brain with a nonbiological equivalent is essentially identical to the following sequence: (i) scan Ray and reinstantiate Ray's mind file into new (nonbiological) Ray, and, then (ii) terminate old Ray. But we concluded above that in such a scenario new Ray is not the same as old Ray. And if old Ray is terminated, well then that's the end of Ray. So the gradual replacement scenario essentially results in the same result: New Ray has been created, and old Ray has been terminated, even if we never saw him missing. So what appears to be the continuing existence of just one Ray is really the creation of new Ray and the end of old Ray.

On yet another hand (we're running out of philosophical hands here), the gradual replacement scenario is not altogether different from what happens normally to our biological selves, in that our particles are always rapidly being replaced. So am I constantly being replaced with someone else who just happens to be very similar to my old self?

I am trying to illustrate why consciousness is not an easy issue. If we talk about consciousness as just a certain type of intelligent skill: the ability to reflect on one's own self and situation, for example, then the issue is not difficult at all because any skill or capability or form of intelligence that one cares to define will be replicated in nonbiological entities (i.e., machines) within a few decades. With this type of *objective* view of consciousness, the conundrums do go away. But a fully objective view does not penetrate to the core of the issue, because the essence of consciousness is *subjective* experience, not objective correlates of that experience.

Will these future machines be capable of having spiritual experiences?

They certainly will claim to. They will claim to be people, and to have the full range of emotional and spiritual experiences that people claim to have. And these will not be idle claims; they will evidence the sort of rich, complex, and subtle behavior one associates with these feelings. How do the claims and behaviors—compelling as they will be—relate to the subjective experience of these reinstantiated people? We keep coming back to the very real but ultimately unmeasurable issue of consciousness.

People often talk about consciousness as if it were a clear property of an entity that can readily be identified, detected, and gauged. If there is one crucial insight that we can make regarding why the issue of consciousness is so contentious, it is the following:

There exists no objective test that can absolutely determine its presence.

Science is about objective measurement and logical implications therefrom, but the very nature of objectivity is that you cannot measure subjective experience—you can only measure correlates of it, such as behavior (and by behavior, I include the actions of components of an entity, such as neurons). This limitation has to do with the very nature of the concepts "objective" and "subjective." Fundamentally, we cannot penetrate the subjective experience of another entity with direct objective measurement. We can certainly make arguments about it: i.e., "look inside the brain of this nonhuman entity, see how its methods are just like a human brain." Or, "see how its behavior is just like human behavior." But in the end, these remain just arguments. No matter how convincing the behavior of a reinstantiated person, some observers will refuse to accept the consciousness of an entity unless it squirts neurotransmitters, or is based on DNA-guided protein synthesis, or has some other specific biologically human attribute.

We assume that other humans are conscious, but that is still an assumption, and there is no consensus amongst humans about the consciousness of nonhuman entities, such as other higher non-human animals. The issue will be even more contentious with regard to future nonbiological entities with human-like behavior and intelligence.

From a practical perspective, we'll accept their claims. Keep in mind that nonbiological entities in the twenty-first century will be extremely intelligent, so they'll be able to convince us that they are conscious. They'll have all the subtle cues that convince us today that humans are conscious. They will be able to make us laugh and cry. And they'll get mad if we don't accept their claims. But this is a political prediction, not a philosophical argument.

On Tubules and Quantum Computing

Over the past several years, Roger Penrose, a noted physicist and philosopher, has suggested that fine structures in the neurons called tubules perform an exotic form of computation called "quantum computing." Quantum computing is computing using what are known as "qu bits," which take on all possible combinations of solutions simultaneously. It can be considered to be an extreme form of parallel processing (because every combination of values of the qu bits is tested simultaneously). Penrose suggests that the tubules and their quantum computing capabilities complicate the concept of recreating neurons and reinstantiating mind files.

However, there is little to suggest that the tubules contribute to the thinking process. Even generous models of human knowledge and capability are more than accounted for by current estimates of brain size, based on contemporary models of neuron functioning that do not include tubules. In fact, even with these tubule-less models, it appears that the brain is conservatively designed with many more connections (by several orders of magnitude) than it needs for its capabilities and capacity. Recent experiments (e.g., the San Diego Institute for Nonlinear Science experiments) showing that hybrid biological-nonbiological networks perform similarly to all biological networks, while not definitive, are strongly suggestive that our tubule-less models of neuron functioning are adequate.

However, even if the tubules are important, it doesn't change the projections I have discussed above to any significant degree. According to my model of computational growth, if the tubules multiplied neuron complexity by a factor of a thousand (and keep in mind that our current tubule-less neuron models are already complex, including on the order of a thousand connections per neuron, multiple nonlinearities and other details), this would delay our reaching brain capacity by only about nine years. If we're off by a factor of a million, that's still only a delay of 17 years. A factor of a billion is around 24 years (keep in mind computation is growing by a double exponential).

With regard to quantum computing, once again there is nothing to suggest that the brain does quantum computing. Just because quantum technology may be feasible does not suggest that the brain is capable of it. We don't have lasers or even radios in our brains either. No one has suggested human capabilities that would require a capacity for quantum computing.

However, even if the brain does do quantum computing, this does not significantly change the outlook for human-level computing (and beyond) nor does it suggest that brain downloading is infeasible. First of all, if the brain does do quantum computing this would only verify that quantum computing is feasible. There would be nothing in such a finding to suggest that quantum computing is restricted to biological mechanisms. Biological quantum computing mechanisms, if they exist, could be replicated. Indeed, recent experiments with small-scale quantum computers appear to be successful.

Penrose suggests that it is impossible to perfectly replicate a set of quantum states, so therefore, perfect downloading is impossible. Well, how perfect does a download have to be? I am at this moment in a very different quantum state (and different in non-quantum ways as well) than I was a minute ago (certainly in a different state than I was before I wrote this paragraph). If we develop downloading technology to the point where the "copies" are as close to the original as the original person changes anyway in the course of one minute, that would be good enough for any conceivable purpose, yet does not require copying quantum states. As the technology improves, the accuracy of the copy could become as close as the original changes within ever-briefer periods of time (e.g., one second, one millisecond).

When it was pointed out to Penrose that neurons (and even neural connections) were too big for quantum computing, he came up with the tubule theory as a possible mechanism for neural quantum computing. So the concerns with quantum computing and tubules have been introduced together. If one is searching for barriers to replicating brain function, it is an ingenious theory, but it fails to introduce any genuine barriers. There is no evidence for it, and even if true, it only delays matters by a decade or two. There is no reason to believe that biological mechanisms (including quantum computing) are inherently impossible to replicate using nonbiological materials and mechanisms. Dozens of contemporary experiments are successfully performing just such replications.

The Noninvasive Surgery-Free Reversible Programmable Distributed Brain Implant

How will we apply technology that is more intelligent than its creators? One might be tempted to respond "Carefully!" But let's take a look at some examples.

Consider several examples of the nanobot technology, which, based on miniaturization and cost reduction trends, will be feasible within 30 years. In addition to scanning your brain, the nanobots will also be able to expand your brain.

Nanobot technology will provide fully immersive, totally convincing virtual reality in the following way. The nanobots take up positions in close physical proximity to every interneuronal connection coming from all of our senses (e.g., eyes, ears, skin). We already have the technology for electronic devices to communicate with neurons in both directions that requires no direct physical contact with the neurons. For example, scientists at the Max Planck Institute have developed "neuron transistors" that can detect the firing of a nearby neuron, or alternatively, can cause a nearby neuron to fire, or suppress it from firing. This amounts to two-way communication between neurons and the electronic-based neuron transistors. The Institute scientists demonstrated their invention by controlling the movement of a living leech from their computer. Again, the primary aspect of nanobot-based virtual reality that is not yet feasible is size and cost.

When we want to experience real reality, the nanobots just stay in position (in the capillaries) and do nothing. If we want to enter virtual reality, they suppress all of the inputs coming from the real senses, and replace them with the signals that would be appropriate for the virtual environment. You (i.e., your brain) could decide to cause your muscles and limbs to move as you normally would, but the nanobots again intercept these interneuronal signals, suppress your real limbs from moving, and instead cause your virtual limbs to move and provide the appropriate movement and reorientation in the virtual environment.

The web will provide a panoply of virtual environments to explore. Some will be recreations of real places; others will be fanciful environments that have no "real" counterpart. Some indeed would be impossible in the physical world (perhaps, because they violate the laws of physics). We will be able to "go" to these virtual environments by ourselves, or we will meet other people there, both real people and simulated people. Of course, ultimately there won't be a clear distinction between the two.

Nanobot technology will be able to expand our minds in virtually any imaginable way. Our brains today are relatively fixed in design. Although we do add patterns of interneuronal connections and neurotransmitter concentrations as a normal part of the learning process, the current overall capacity of the human brain is highly constrained, restricted to a mere hundred trillion connections. Brain implants based on massively distributed intelligent nanobots will ultimately expand our memories a trillion fold, and otherwise vastly improve all of our sensory, pattern recognition and cognitive abilities. Since the nanobots are communicating with each other over a wireless local area network, they can create any set of new neural connections, can break existing connections (by suppressing neural firing), can create new hybrid biological-nonbiological networks as well as adding vast new nonbiological networks.

Using nanobots as brain extenders is a significant improvement over the idea of surgically installed neural implants, which are beginning to be used today. Nanobots will be introduced without surgery, essentially just by injecting or even swallowing them. They can all be directed to leave, so the process is easily reversible. They are programmable, in that they can provide virtual reality one minute, and a variety of brain extensions the next. They can change their configuration, and clearly can alter their software. Perhaps most importantly, they are massively distributed and therefore can take up billions or trillions of positions throughout the brain, whereas a surgically introduced neural implant can only be placed in one or at most a few locations.

A Clear and Future Danger

Technology has always been a double-edged sword, bringing us longer and healthier life spans, freedom from physical and mental drudgery, and many new creative possibilities on the one hand, while introducing new and salient dangers on the other. We still live today with sufficient nuclear weapons (not all of which appear to be well accounted for) to end all mammalian life on the planet. Bioengineering is in the early stages of enormous strides in reversing disease and aging processes. However, the means and knowledge exist in a routine college bioengineering lab to create unfriendly pathogens more dangerous than nuclear weapons. For the twenty-first century, we will see the same intertwined potentials: a great feast of creativity resulting from human intelligence expanded a trillion-fold combined with many grave new dangers.

Consider unrestrained nanobot replication. Nanobot technology requires billions or trillions of such intelligent devices to be useful. The most cost-effective way to scale up to such levels is through self-replication, essentially the same approach used in the biological world. And in the same way that biological self-replication gone awry (i.e., cancer) results in biological destruction, a defect in the mechanism curtailing nanobot self-replication would endanger all physical entities, biological or otherwise.

Other salient concerns include "who is controlling the nanobots?" and "who are the nanobots talking to?" Organizations (e.g., governments, extremist groups) or just a clever individual could put trillions of undetectable nanobots in the water or food supply of an individual or of an entire population. These "spy" nanobots could then monitor, influence, and even control our thoughts and actions. In addition to introducing physical spy nanobots, existing nanobots could be influenced through software viruses and other software "hacking" techniques.

My own expectation is that the creative and constructive applications of this technology will dominate, as I believe they do today. But there will be a valuable (and increasingly vocal) role for a concerned and constructive Luddite movement (i.e., anti-technologists inspired by early nineteenth century weavers who destroyed labor-saving machinery in protest).

Living Forever

Once brain porting technology has been refined and fully developed, will this enable us to live forever? The answer depends on what we mean by living and dying. Consider what we do today with our personal computer files. When we change from one personal computer to a less obsolete model, we don't throw all our files away; rather we copy them over to the new hardware. Although our software files do not necessary continue their existence forever, the longevity of our personal computer software is completely separate and disconnected from the hardware that it runs on. When it comes to our personal mind file, however, when our human hardware crashes, the software of our lives dies with it. However, this will not continue to be the case when we have the means to store and restore the thousands of trillions of bytes of information stored and represented in our brains.

The longevity of one's mind file will not be dependent, therefore, on the continued viability of any particular hardware medium. Ultimately software-based humans, albeit vastly extended beyond the severe limitations of humans as we know them today, will live out on the web, projecting bodies whenever they need or want them, including virtual bodies in diverse realms of virtual reality, holographically projected bodies, and physical bodies comprised of nanobot swarms, and other forms of nanotechnology.

A software-based human will be free, therefore, from the constraints of any particular thinking medium. Today, we are each confined to a mere hundred trillion connections, but humans at the end of the twenty-first century can grow their thinking and thoughts without limit. We may regard this as a form of immortality, although it is worth pointing out that data and information do not necessarily last forever. Although not dependent on the viability of the hardware it runs on, the longevity of information depends on its relevance, utility, and accessibility. If you've ever tried to retrieve information from an obsolete form of data storage in an old obscure format (e.g., a reel of magnetic tape from a 1970 minicomputer), you will understand the challenges in keeping software viable. However, if we are diligent in maintaining our mind file, keeping current backups, and porting to current formats and mediums, then a form of immortality can be attained, at least for software-based humans. Our mind file—our personality, skills, memories—all of that is lost today when our biological hardware crashes. When we can access, store, and restore that information, then its longevity will no longer be tied to hardware permanence.

Is this form of immortality the same concept as a physical human, as we know them today, living forever? In one sense it is, because as I pointed out earlier, we are not a constant collection of matter. Only our pattern of matter and energy persists, and even that gradually changes. Similarly, it will be the pattern of a software human that persists and develops and changes gradually.

But is that person based on my mind file, who migrates across many computational substrates, and who outlives any particular thinking medium, really me? We come back to the same questions of consciousness and identity, issues that have been debated since the Platonic dialogues. As we go through the twenty-first century, these will not remain polite philosophical debates, but will be confronted as vital and practical issues.

A related question is, "Is death desirable?" A great deal of our effort goes into avoiding it. We make extraordinary efforts to delay it, and indeed often consider its intrusion a tragic event. Yet we might find it hard to live without it. We consider death as giving meaning to our lives. It gives importance and value to time. Time could become meaningless if there were too much of it.

The Next Step in Evolution

But I regard the freeing of the human mind from its severe physical limitations of scope and duration as the necessary next step in evolution. Evolution, in my view, represents the purpose of life. That is, the purpose of life—and of our lives—is to evolve.

What does it mean to evolve? Evolution moves towards greater complexity, greater elegance, greater knowledge, greater intelligence, greater beauty, greater creativity, greater love. And God has been called all these things, only without any limitation: infinite knowledge, infinite intelligence, infinite beauty, infinite creativity, and infinite love. Evolution does not achieve an infinite level, but as it explodes exponentially, it certainly moves in that direction. So evolution moves inexorably towards our conception of God, albeit never reaching this ideal. Thus the freeing of our thinking from the severe limitations of its biological form may be regarded as an essential spiritual quest.

In making this statement, it is important to emphasize that terms like evolution, destiny, and spiritual quest are observations about the end result, not justifications for it. I am not saying that technology will evolve to human levels and beyond simply because it is our destiny and the satisfaction of a spiritual quest. Rather my projections result from a methodology based the dynamics underlying the (double) exponential growth of technological processes. The primary force driving technology is economic imperative. We are moving towards machines with human level intelligence (and beyond) as the result of millions of advances, each with their own economic justification. To use an example from my own experience at one of my companies (Kurzweil Applied Intelligence), whenever we came up with a slightly more intelligent version of speech recognition, the new version invariably had greater value than the earlier generation and, as a result, sales increased. It is interesting to note that in the example of speech recognition software, the three primary surviving competitors (Kurzweil—now Lernout & Hauspie, Dragon, and IBM) stayed very close to each other in the intelligence of their software. A few other companies that failed to do so (e.g., Speech Systems) went out of business. At any point in time, we would be able to sell the version prior to the latest version for perhaps a quarter of the price of the current version. As for versions of our technology that were two generations old, we couldn't even give those away. This phenomenon is not only true for pattern recognition and other "AI" software. It's true of any product from cars to bread makers. And if the product itself doesn't exhibit some level of intelligence, then intelligence in the manufacturing and marketing methods have a major effect on the success and profitability of an enterprise.

There is a vital economic imperative to create more intelligent technology. Intelligent machines have enormous value. That is why they are being built. There are tens of thousands of projects that are advancing intelligent machines in many diverse ways. The support for "high tech" in the business community (mostly software) has grown enormously. When I started my OCR and speech synthesis company in 1974, the total U.S. annual venture capital investment in high tech was around $8 million. Total high tech IPOs for 1974 was about the same figure. Today, high tech IPOs (principally software) are about $30 million per day, more than a thousand fold increase.

We will continue to build more powerful computational mechanisms because it creates enormous value. We will reverse engineer the human brain not because it is our destiny, but because there is valuable information to be found there that will provide insights in building more intelligent (and more valuable) machines. We would have to repeal capitalism and every visage of economic competition to stop this progression.

By the second half of this next century, there will be no clear distinction between human and machine intelligence. On the one hand, we will have biological brains vastly expanded through distributed nanobot-based implants. On the other hand, we will have fully nonbiological brains that are copies of human brains, albeit also vastly extended. And we will have myriad other varieties of intimate connection between human thinking and the technology it has fostered.

Ultimately, nonbiological intelligence will dominate because it is growing at a double exponential rate, whereas for all practical purposes biological intelligence is at a standstill. By the end of the twenty-first century, nonbiological thinking will be trillions of trillions of times more powerful than that of its biological progenitors, although still of human origin. It will continue to be the human-machine civilization taking the next step in evolution.

Before the next century is over, the Earth's technology-creating species will merge with its computational technology. After all, what is the difference between a human brain enhanced a trillion fold by nanobot-based implants, and a computer whose design is based on high resolution scans of the human brain, and then extended a trillion-fold?

Most forecasts of the future seem to ignore the revolutionary impact of the inevitable emergence of computers that match and ultimately vastly exceed the capabilities of the human brain, a development that will be no less important than the evolution of human intelligence itself some thousands of centuries ago.

2

I Married a Computer

John Searle

11.9.0Y

Kurzweil's Central Argument

Moore's Law on Integrated Circuits was first formulated by Gordon Moore, former head of Intel, in the mid-Sixties. I have seen different versions of it, but the basic idea is that better chip technology will produce an exponential increase in computer power. Every two years you get twice as much computer power and capacity for the same amount of money. Anybody who, like me, buys a new computer every few years observes Moore's Law in action. Each time I buy a new computer I pay about the same amount of money as, and sometimes even less than, I paid for the last computer, but I get a much more powerful machine. And according to

John Searle is Mills Professor of the Philosophy of Mind at the University of California at Berkeley and author of many books including Rationality in Action *(Cambridge: MIT Press, 2001).*

Ray Kurzweil, who is himself a distinguished software engineer and inventor, "There have been about thirty-two doublings of speed and capacity since the first operating computers were built in the 1940s."

Furthermore, we can continue to project this curve of increased computing power into the indefinite future. Moore's Law itself is about chip technology, and Kurzweil tells us that this technology will reach an upper limit when we reach the theoretical possibilities of the physics of silicon in about the year 2020. But Kurzweil tells us not to worry, because we know from evolution that some other technology will take over and "pick up where Moore's Law will have left off, without missing a beat." We know this, Kurzweil assures us, from "The Law of Accelerating Returns," which is a basic attribute of the universe; indeed it is a sublaw of "The Law of Time and Chaos." These last two laws are Kurzweil's inventions.

It is fair to say that *The Age of Spiritual Machines* is an extended reflection on the implications of Moore's Law, and is a continuation of a line of argument begun in his earlier book, *The Age of Intelligent Machines*. He begins by placing the evolution of computer technology within the context of evolution in general, and he places that within the history of the universe. The book ends with a brief history of the universe, which he calls "Time Line," beginning at the Big Bang and going to 2099.

So what, according to Kurzweil and Moore's Law, does the future hold for us? We will very soon have computers that vastly exceed us in intelligence. Why does increase in computing power automatically generate increased intelligence? Because intelligence, according to Kurzweil, is a matter of getting the right formulas in the right combination and then applying them over and over, in his sense "recursively," until the problem is solved. With sheer computational brute force, he thinks, you can solve any solvable problem. It is true, Kurzweil admits, that computational brute force is not enough by itself, and ultimately you will need "the complete set of unifying formulas that underlie intelligence." But we are well on the way to discovering these formulas: "Evolution determined an answer to this problem in a few billion years. We've made a good start in a few thousand years. We are likely to finish the job in a few more decades."

Let us suppose for the sake of argument that we soon will have computers that are more "intelligent" than we are. Then what? This is where Kurzweil's book begins to go over the edge. First off, according to him, living in this slow, wet, messy hardware of our own neurons may be sentimentally appealing, like living in an old shack with a view of the ocean, but within a very few decades, sensible people will get out of neurons and have themselves "downloaded" onto some decent hardware. How is this to be done? You will have your entire brain and nervous system scanned, and then, when you and the experts you consult have figured out the programs exactly, you reprogram an electronic circuit with your programs and database. The electronic circuit will have more "capacity, speed, and reliability" than neurons. Furthermore, when the parts wear out they permit much easier replacement than neurons do.

So that is the first step. You are no longer locked into wet, slow, messy, and above all decaying hardware; you are upgraded into the latest circuitry. But it would be no fun just to spend life as a desktop in the office, so you will need a new body. And how is that to be done? Nanotechnology, the technology of building objects atom by atom and molecule by molecule, comes to the rescue. You replace your old body atom by atom. "We will be able to reconstruct any or all of our bodily organs and systems, and do so at the cellular level. ...We will then be able to grow stronger, more capable organs by redesigning the cells that constitute them and building them with far more versatile and durable materials." Kurzweil does not tell us anything at all about what these materials might be, but they clearly will not be flesh and blood, calcium bones and nucleoproteins.

Evolution will no longer occur in organic carbon-based materials but will pass to better stuff. However, though evolution will continue, we as individuals will no longer suffer from mortality. Even if you do something stupid like get blown up, you still keep a replacement copy of your programs and database on the shelf so you can be completely reconstructed at will. Furthermore, you can change your whole appearance and other characteristics at will, "in a split second." You can look like Marlon Brando one minute and like Marlene Dietrich the next.

In Kurzweil's vision, there is no conflict between human beings and machines, because we will all soon, within the lifetimes of most people alive today, become machines. Strictly speaking we will become software. As he puts it, *"We will be software, not hardware"* (italics his) and can inhabit whatever hardware we like best. There will not be any difference between robots and us. "What, after all, is the difference between a human who has upgraded her body and brain using new nanotechnology, and computational technologies and a robot who has gained an intelligence and sensuality surpassing her human creators?" What, indeed? Among the many advantages of this new existence is that you will be able to read any book in just a few seconds. You could read Dante's *Divine Comedy* in less time than it takes to brush your teeth.

Kurzweil recognizes that there are some puzzling features of this utopian dream. If I have my programs downloaded onto a better brain and hardware but leave my old body still alive, which one is really me? The new robot or the old pile of junk? A problem he does not face: Suppose I make a thousand or a million copies of myself. Are they all me? Who gets to vote? Who owns my house? Who is my spouse married to? Whose driver's license is it, anyhow?

What will sex life be like in this brave new world? Kurzweil offers extended, one might even say loving, accounts. His main idea is that virtual sex will be just as good as, and in many ways better than, old-fashioned sex with real bodies. In virtual sex your computer brain will be stimulated directly with the appropriate signal without the necessity of any other human body, or even your own body. Here is a typical passage:

> Virtual touch has already been introduced, but the all-enveloping, highly realistic, visual-auditory-tactile virtual environment will not be perfected until the second decade of the twenty-first century. At this point,

virtual sex becomes a viable competitor to the real
thing. Couples will be able to engage in virtual sex
regardless of their physical proximity. Even when
proximate, virtual sex will be better in some ways and
certainly safer. Virtual sex will provide sensations that
are more intense and pleasurable than conventional
sex, as well as physical experiences that currently do
not exist.

The section on prostitution is a little puzzling to me:

Prostitution will be free of health risks, as will virtual
sex in general. Using wireless, very-high-bandwidth
communication technologies, neither sex workers nor
their patrons need to leave their homes.

But why pay, if it is all an electrically generated fantasy anyway?
Kurzweil seems to concede as much when he says, "Sex workers
will have competition from simulated—computer generated—partners." And, he goes on, "once the simulated virtual partner is as capable, sensual, and responsive as a real human virtual partner, who's
to say that the simulated virtual partner isn't a real, albeit virtual,
person?"

It is important to emphasize that all of this is seriously intended.
Kurzweil does not think he is writing a work of science fiction, or a
parody or satire. He is making serious claims that he thinks are based
on solid scientific results. He is himself a distinguished computer
scientist and inventor and so can speak with some authority about
current technology. One of his rhetorical strategies is to cite earlier
successful predictions he has made as evidence that the current ones
are likely to come true as well. Thus he predicted within a year when
a computer chess machine would be able to beat the world chess
champion, and he wants us to take his prediction that we will all
have artificial brains within a few decades as just more of the same
sort of solidly based prediction. Because he frequently cites the IBM

chess-playing computer Deep Blue as evidence of superior intelligence in the computer, it is worth examining its significance in more detail.

Deep Blue and the Chinese Room

When it was first announced that Deep Blue had beaten Gary Kasparov, the media gave it a great deal of attention, and I suspect that the attitude of the general public was that what was going on inside Deep Blue was much the same sort of thing as what was going on inside Kasparov, only Deep Blue was better at that sort of thing and was doing a better job. This reveals a total misunderstanding of computers, and the programmers, to their discredit, did nothing to remove the misunderstanding. Here is the difference: Kasparov was consciously looking at a chessboard, studying the position and trying to figure out his next move. He was also planning his overall strategy and no doubt having peripheral thoughts about earlier matches, the significance of victory and defeat, etc. We can reasonably suppose he had all sorts of unconscious thoughts along the same lines. Kasparov was, quite literally, playing chess. None of this whatever happened inside Deep Blue. Nothing remotely like it.

Here is what happened inside Deep Blue. The computer has a bunch of meaningless symbols that the programmers use to represent the positions of the pieces on the board. It has a bunch of equally meaningless symbols that the programmers use to represent options for possible moves. The computer does not know that the symbols represent chess pieces and chess moves, because it does not know anything. As far as the computer is concerned, the symbols could be used to represent baseball plays or dance steps or numbers or nothing at all.

If you are tempted to think that the computer literally understands chess, then remember that you can use a variation on the Chinese Room Argument against the chess-playing computer. Let us call it the Chess Room Argument. Imagine that a man who does not know how to play chess is locked inside a room, and there he is given a set of, to him, meaningless symbols. Unknown to him, these represent positions on a chessboard. He looks up in a book what he is supposed to do, and he passes back more meaningless symbols. We can suppose that if the rule book, i.e., the program, is skillfully written, he will win chess games. People outside the room will say, "This man understands chess, and in fact he is a good chess player because he wins." They will be totally mistaken. The man understands nothing of chess; he is just a computer. And the point of the parable is this: If the man does not understand chess on the basis of running the chess-playing program, neither does any other computer solely on that basis.

The Chinese Room Argument shows that just carrying out the steps in a computer program is not by itself sufficient to guarantee cognition. Imagine that I, who do not know Chinese, am locked in a room with a computer program for answering written questions, put to me in Chinese, by providing Chinese symbols as answers. If properly programmed I will provide answers indistinguishable from those of native Chinese speakers, but I still do not understand Chinese. And if I don't, neither does any other computer solely on the basis of carrying out the program. See my "Minds, Brains and Programs," *Behavioral and Brain Sciences*, Vol. 3 (1980) for the first statement of this argument. See also "The Myth of the Computer," published in the *New York Review of Books*, April 29, 1982.

The ingenuity of the hardware engineers and the programmers who programmed Deep Blue was manifested in this: from the point of view of mathematical game theory, chess is a trivial game because each side has perfect information. You know how many pieces you and your opponent have and what their locations are. You can theoretically know all of your possible moves and all of your opponent's possible countermoves. It is in principle a solvable game. The interest of chess for human beings and the problem for programmers arises from what is called a combinatorial explosion. In chess at any given point there is a finite number of possible moves. Suppose I am white and I have, say, eight possible moves. For each of these moves there is a set of possible countermoves by black and to them a set of possible moves by white, and so on up exponentially. After a few levels the number of possible positions on the board is astronomical and no human being can calculate them all. Indeed, after a few more moves the numbers are so huge that no existing computer can calculate them. At most a good chess player might calculate a few hundred.

This is where Deep Blue had the advantage. Because of the increased computational power of the machinery, it could examine 200 million positions per second; so, according to the press accounts at the time, the programmers could program the machine to follow out the possibilities to twelve levels: first white, then black, then white, and so on to the twelfth power. For some positions the machine could calculate as far as forty moves ahead. Where the human player can imagine a few hundred possible positions, the computer can scan billions.

But what does it do when it has finished scanning all these positions? Here is where the programmers have to exercise some judgment. They have to design a "scoring function." The machine attaches a numerical value to each of the final positions of each of the possible paths that developed in response to each of the initial moves. So for example a situation in which I lose my queen has a low number, a position in which I take your queen has a high number. Other factors are taken into consideration in determining the number: the mobility of the pieces (how many moves are available), the position of the pawns, etc. IBM experts are very secretive about the details of their scoring function, but they claim to use about 8,000 factors. Then, once the machine has assigned a number to all the final positions, it assigns numbers to the earlier positions leading to the final positions depending on the numbers of those final positions. The machine then selects the symbol that represents the move that leads to the highest number. It is that simple and that mechanical, though it involves a lot of symbol shuffling to get there. The real competition was not between Kasparov and the machine, but between Kasparov and a team of engineers and programmers.

Kurzweil assures us that Deep Blue was actually thinking. Indeed he suggests that it was doing more thinking than Kasparov. But what was it thinking about? Certainly not about chess, because it had no way of knowing that these symbols represent chess positions. Was it perhaps thinking about numbers? Even that is not true, because it had no way of knowing that the symbols assigned represented numerical values. The symbols in the computer mean nothing at all to the computer. They mean something to us because we have built and programmed the computer so that it can manipulate symbols in a way that is meaningful to us. In this case we are using the computer symbols to represent chess positions and chess moves.

Now, with all this in mind, what psychological or philosophical significance should we attach to Deep Blue? It is, of course, a wonderful hardware and software achievement of the engineers and the programmers, but as far as its relevance to human psychology is concerned, it seems to me of no interest whatsoever. Its relevance is similar to that of a pocket calculator for understanding human thought processes when doing arithmetic. I was frequently asked by reporters at the time of the triumph of Deep Blue if I did not think that this was somehow a blow to human dignity. I think it is nothing of the sort. Any pocket calculator can beat any human mathematician at arithmetic. Is this a blow to human dignity? No, it is rather a credit to the ingenuity of programmers and engineers. It is simply a result of the fact that we have a technology that enables us to build tools to do things that we cannot do, or cannot do as well or as fast, without the tools.

Kurzweil also predicts that the fact that a machine can beat a human being in chess will lead people to say that chess was not really important anyway. But I do not see why. Like all games, chess is built around the human brain and body and its various capacities and limitations. The fact that Deep Blue can go through a series of electrical processes that we can interpret as "beating the world champion at chess" is no more significant for human chess playing than it would be significant for human football playing if we built a steel robot which could carry the ball in a way that made it impossible for the robot to be tackled by human beings. The Deep Blue chess player is as irrelevant to human concerns as is the Deep Blue running back.

Some Conceptual Confusions

I believe that Kurzweil's book exhibits a series of conceptual confusions. These are not all Kurzweil's fault; they are common to the prevailing culture of information technology, and especially to the subculture of artificial intelligence, of which he is a part. Much of the confusion in this entire field derives from the fact that people on both sides of the debate tend to suppose that what is at issue is the success or failure of computational simulations. Are human beings "superior" to computers or are computers superior? That is not the point at issue at all. The question is not whether computers can succeed at doing this or that. For the sake of argument, I am just going to assume that everything Kurzweil says about the increase in computational power is true. I will assume that computers both can and will do everything he says they can and will do, that there is no question about the capacity of human designers and programmers to build ever faster and more powerful pieces of computational machinery. My point is that to the issues that really concern us about human consciousness and cognition, these successes are irrelevant.

What, then, is at issue? Kurzweil's book exhibits two sets of confusions, which I shall consider in order.

(1) He confuses the computer simulation of a phenomenon with a duplication or re-creation of that phenomenon. This comes out most obviously in the case of consciousness. Anybody who is seriously considering having his "program and database" downloaded onto some hardware ought to wonder whether or not the resulting hardware is going to be conscious. Kurzweil is aware of this problem, and he keeps coming back to it at various points in his book. But his attempt to solve the problem can only be said to be plaintive. He does not claim to know that machines will be conscious, but he insists that they will claim to be conscious, and will continue to engage in discussions about whether they are conscious, and consequently their claims will be largely accepted. People will eventually just come to accept without question that machines are conscious.

But this misses the point. I can already program my computer so that it says that it is conscious—i.e., it prints out "I am conscious"— and a good programmer can even program it so that it will carry on a rudimentary argument to the effect that it is conscious. But that has nothing to do with whether or not it really is conscious. Actual human brains cause consciousness by a series of specific neurobiological processes in the brain. What the computer does is a simulation of these processes, a symbolic model of the processes. But the computer simulation of brain processes that produce consciousness stands to real consciousness as the computer simulation of the stomach processes that produce digestion stands to real digestion. You do not cause digestion by doing a computer simulation of digestion. Nobody thinks that if we had the perfect computer simulation running on the computer, we could stuff a pizza into the computer and it would thereby digest it. It is the same mistake to suppose that when a computer simulates the processes of a conscious brain it is thereby conscious.

The computer, as we saw in our discussion of the chess-playing program, succeeds by manipulating formal symbols. The symbols themselves are quite meaningless; they have only the meaning we have attached to them. The computer knows nothing of this, it just shuffles the symbols. And those symbols are not by themselves sufficient to guarantee equivalent causal powers to actual biological machinery like human stomachs and human brains.

Kurzweil points out that not all computers manipulate symbols. Some recent machines simulate the brain by using networks of parallel processors called "neural nets," which try to imitate certain features of the brain. But that is no help. We know from the Church-Turing Thesis, a mathematical result, that any computation that can be carried out on a neural net can be carried out on a symbol-manipulating machine. The neural net gives no increase in computational power. And simulation is still not duplication.

But, someone is bound to ask, can you prove that the computer is not conscious? The answer to this question is: Of course not. I cannot prove that the computer is not conscious, any more than I can prove that the chair I am sitting on is not conscious. But that is not the point. It is out of the question, for purely neurobiological reasons, to suppose that the chair or the computer is conscious. The point for the present discussion is that the computer is not designed to be conscious. It is designed to manipulate symbols in a way that carries out the steps in an algorithm. It is not designed to duplicate the actual causal powers of the brain to cause consciousness. It is designed to enable us to simulate any process that we can describe precisely.

Kurzweil is aware of this objection and tries to meet it with a slippery-slope argument: We already have brain implants, such as cochlear implants in the auditory system, that can duplicate and not merely simulate certain brain functions. What is to prevent us from a gradual replacement of all the brain anatomy that would preserve and not merely simulate our consciousness and the rest of our mental life? In answer to this, I would point out that he is now abandoning the main thesis of the book, which is that what is important for consciousness and other mental functions is entirely a matter of computation. In his words, we will become software, not hardware.

I believe that there is no objection in principle to constructing an artificial hardware system that would duplicate the powers of the brain to cause consciousness using some chemistry different from neurons. But to produce consciousness any such system would have to duplicate the actual causal powers of the brain. And we know, from the Chinese Room Argument, that computation by itself is insufficient to guarantee any such causal powers, because computation is defined entirely in terms of the manipulation of abstract formal symbols.

(2) The confusion between simulation and duplication is a symptom of an even deeper confusion in Kurzweil's book, and that is between those features of the world that exist intrinsically, or independently of human observation and conscious attitudes, and those features of the world that are dependent on human attitudes—the distinction, in short, between features that are observer-independent and those that are observer-relative.

Examples of observer-independent features are the sorts of things discussed in physics and chemistry. Molecules, and mountains, and tectonic plates, as well as force, mass, and gravitational attraction, are all observer-independent. Since relativity theory we recognize that some of their limits are fixed by reference to other systems, but none of them are observer-dependent in the sense of requiring the thoughts of conscious agents for their existence. On the other hand, such features of the world as money, property, marriage, government, and football games require conscious observers and agents in order for them to exist as such. A piece of paper has intrinsic or observer-independent chemical properties, but a piece of paper is a dollar bill only in a way that is observer-dependent or observer-relative.

In Kurzweil's book many of his crucial notions oscillate between having a sense that is observer-independent, and another sense that is observer-relative. The two most important notions in the book are intelligence and computation, and both of these exhibit precisely this ambiguity. Take intelligence first.

In a psychological, observer-independent sense, I am more intelligent than my dog, because I can have certain sorts of mental processes that he cannot have, and I can use these mental capacities to solve problems that he cannot solve. But in this psychological sense of intelligence, wristwatches, pocket calculators, computers, and cars are not candidates for intelligence, because they have no mental life whatever.

In an observer-relative sense, we can indeed say that lots of machines are more intelligent than human beings because we have designed the machines in such a way as to help us solve problems that we cannot solve, or cannot solve as efficiently, in an unaided fashion. Chess-playing machines and pocket calculators are good examples. Is the chess-playing machine really more intelligent at chess than Kasparov? Is my pocket calculator more intelligent than I at arithmetic? Well, in an intrinsic or observer-independent sense, of course not, the machine has no intelligence whatever, it is just an electronic circuit that we have designed, and can ourselves operate, for certain purposes. But in the metaphorical or observer-relative sense, it is perfectly legitimate to say that the chess-playing machine has more intelligence, because it can produce better results. And the same can be said for the pocket calculator.

There is nothing wrong with using the word "intelligence" in both senses, provided you understand the difference between the observer-relative and the observer-independent. The difficulty is that this word has been used as if it were a scientific term, with a scientifically precise meaning. Indeed, many of the exaggerated claims made on behalf of "artificial intelligence" have been based on this systematic confusion between observer-independent, psychologically relevant intelligence and metaphorical, observer-relative, psychologically irrelevant ascriptions of intelligence. There is nothing wrong with the metaphor as such; the only mistake is to think that it is a scientifically precise and unambiguous term. A better term than "artificial intelligence" would have been "simulated cognition."

Exactly the same confusion occurs over the notion of "computation." There is a literal sense in which human beings are computers because, for example, we can compute 2+2=4. But when we design a piece of machinery to carry out that computation, the computation 2+2=4 exists only relative to our assignment of a computational interpretation to the machine. Intrinsically, the machine is just an electronic circuit with very rapid changes between such things as voltage levels. The machine knows nothing about arithmetic just as it knows nothing about chess. And it knows nothing about computation either, because it knows nothing at all. We use the machinery to compute with, but that does not mean that the computation is intrinsic to the physics of the machinery. The computation is observer-relative, or to put it more traditionally, "in the eye of the beholder."

This distinction is fatal to Kurzweil's entire argument, because it rests on the assumption that the main thing humans do in their lives is compute. Hence, on his view, if—thanks to Moore's Law—we can create machines that can compute better than humans, we have equaled and surpassed humans in all that is distinctively human. But in fact humans do rather little that is literally computing. Very little of our time is spent working out algorithms to figure out answers to questions. Some brain processes can be usefully described as if they were computational, but that is observer-relative. That is like the attribution of computation to commercial machinery, in that it requires an outside observer or interpreter.

Another result of this confusion is a failure on Kurzweil's part to appreciate the significance of current technology. He describes the use of strands of DNA to solve the Traveling Salesman Problem— the problem of how to plot a route for a salesman so that he never goes through the same city twice—as if it were the same sort of thing as the use, in some cases, of neural implants to cure Parkinson's Disease. But the two cases are completely different. The cure for Parkinson's Disease is an actual, observer-independent causal effect on the human brain. But the sense in which the DNA strands stand for or represent different cities is entirely observer-relative. The DNA knows nothing about cities.

It is worth pointing out here that when Alan Turing first invented the idea of the computer, the word "computer" meant "person who computes." "Computer" was like "runner" or "skier." But as commercial computers have become such an essential part of our lives, the word "computer" has shifted in meaning to mean "machinery designed by us to use for computing," and, for all I know, we may go through a change of meaning so that people will be said to be computers only in a metaphorical sense. It does not matter as long as you keep the conceptual distinction clear between what is intrinsically going on in the machinery, however you want to describe it, and what is going on in the conscious thought processes of human beings. Kurzweil's book fails throughout to perceive these distinctions.

The Problem of Consciousness

We are now in the midst of a technological revolution that is full of surprises. No one thirty years ago was aware that one day household computers would become as common as dishwashers. And those of us who used the old Arpanet of twenty years ago had no idea that it would evolve into the Internet. This revolution cries out for interpretation and explanation. Computation and information processing are both harder to understand and more subtle and pervasive in their effects on civilization than were earlier technological revolutions such as those of the automobile and television. The two worst things that experts can do when explaining this technology to the general public are first to give the readers the impression that they understand something they do not understand, and second to give the impression that a theory has been established as true when it has not.

Kurzweil's book suffers from both of these defects. The title of the book is *The Age of Spiritual Machines*. By "spiritual," Kurzweil means conscious, and he says so explicitly. The implications are that if you read his book you will come to understand the machines and that we have overwhelming evidence that they now are or will shortly be conscious. Both of these implications are false. You will not understand computing machinery from reading Kurzweil's book. There is no sustained effort to explain what a computer is and how it works. Indeed one of the most fundamental ideas in the theory of computation, the Church-Turing Thesis, is stated in a way which is false.

Here is what Kurzweil says:

> This thesis says that all problems that a human being can solve can be reduced to a set of algorithms, supporting the idea that machine intelligence and human intelligence are essentially equivalent.

That definition is simply wrong. The actual thesis comes in different formulations (Church's is different from Turing's, for example), but the basic idea is that any problem that has an algorithmic solution can be solved on a Turing machine, a machine that manipulates only two kinds of symbols, the famous zeroes and ones.

Where consciousness is concerned, the weaknesses of the book are even more disquieting. One of its main themes, in some ways the main theme, is that increased computational power gives us good, indeed overwhelming, reason to think we are moving into an era when computing machinery artifacts, machines made by us, will be conscious, "the age of spiritual machines." But from everything we know about the brain, and everything we know about computation, increased computational power in a machine gives us no reason whatever to suppose that the machine is duplicating the specific neurobiological powers of the brain to create consciousness. Increased computer power by itself moves us not one bit closer to creating a conscious machine. It is just irrelevant.

Suppose you took seriously the project of building a conscious machine. How would you go about it? The brain is a machine, a biological machine to be sure, but a machine all the same. So the first step is to figure out how the brain does it and then build an artificial machine that has an equally effective mechanism for causing consciousness. These are the sorts of steps by which we built an artificial heart. The problem is that we have very little idea of how the brain does it. Until we do, we are most unlikely to produce consciousness artificially in nonbiological materials. When it comes to understanding consciousness, ours is not the age of spiritual machines. It is more like the age of neurobiological infancy, and in our struggles to get a mature science of the brain, Moore's Law provides no answers.

A Brief Recapitulation

In response to my initial review of Kurzweil's book in *The New York Review of Books*, Kurzweil wrote both a letter to the editor and a more extended rebuttal on his website. He claims that I presented a "distorted caricature" of his book, but he provided no evidence of any distortion. In fact I tried very hard to be scrupulously accurate both in reporting his claims and in conveying the general tone of futuristic techno-enthusiasm that pervades the book. So at the risk of pedantry, let's recapitulate briefly the theses in his book that I found most striking:

(1) Kurzweil thinks that within a few decades we will be able to download our minds onto computer hardware. We will continue to exist as computer software. *"We will be software, not hardware"* (p. 129, his italics). And "the essence of our identity will switch to the permanence of our software" (p.129).

(2) According to him, we will be able to rebuild our bodies, cell by cell, with different and better materials using "nanotechnology." Eventually, " there won't be a clear difference between humans and robots" (p.148).

(3) We will be immortal, not only because we will be made of better materials, but because even if we were destroyed we will keep copies of our programs and databases in storage and can be reconstructed at will. "Our immortality will be a matter of being sufficiently careful to make frequent back-ups," he says, adding the further caution: "If we're careless about this, we'll have to load an old backup copy and be doomed to repeat our recent past" (p. 129). (What is this supposed to mean? That we will be doomed to repeat our recent car accident and spring vacation?)

(4) We will have overwhelming evidence that computers are conscious. Indeed there will be "no longer any clear distinction between humans and computers" (p. 280).

(5) There will be many advantages to this new existence, but one he stresses is that virtual sex will soon be a "viable competitor to the real thing," affording "sensations that are more intense and pleasurable than conventional sex" (p. 147).

Frankly, had I read this as a summary of some author's claims, I might think it must be a "distorted caricature," but Kurzweil did in fact make each of these claims, as I show by extensive quotation. In his letter he did not challenge me on any of these central points. He conceded by his silence that my understanding of him on these central issues is correct. So where is the "distorted caricature?"

I then point out that his arguments are inadequate to establish any of these spectacular conclusions. They suffer from a persistent confusion between simulating a cognitive process and duplicating it, and even worse confusion between the observer-relative, in-the-eye-of-the-beholder sense of concepts like intelligence, thinking, etc., and the observer-independent intrinsic sense.

What has he to say in response? Well, about the main argument he says nothing. About the distinction between simulation and duplication, he says he is describing neither simulations of mental powers nor re-creations of the real thing, but "functionally equivalent re-creations." But the notion "functionally equivalent" is ambiguous precisely between simulation and duplication. What exactly functions to do exactly what? Does the computer simulation function to enable the system to have *external* behavior, which is *as if* it were conscious, or does it function to actually cause *internal* conscious states? For example, my pocket calculator is "functionally equivalent" to (indeed better than) me in producing answers to arithmetic problems, but it is not thereby functionally equivalent to me in producing the conscious thought processes that go with solving arithmetic problems. Kurzweil's argument about consciousness is based on the assumption that the external behavior is overwhelming evidence for the presence of the internal conscious states. He has no answer to my objection that once you know that the computer works by shuffling symbols, its behavior is no evidence at all for consciousness. The notion of functional equivalence does not overcome the distinction between simulation and duplication; it just disguises it for one step.

In his letter he told us he is interested in doing "reverse engineering" to figure out how the brain works. But in the book there is virtually nothing about the actual working of the brain and how the specific electro-chemical properties of the thalamo-cortical system could produce consciousness. His attention rather is on the computational advantages of superior hardware.

On the subject of consciousness there actually is a "distorted caricature," but it is Kurzweil's distorted caricature of my arguments. He said, "Searle would have us believe that you can't be conscious if you don't squirt neurotransmitters (or some other specific biological process)." Here is what I actually wrote: "I believe there is no objection in principle to constructing an artificial hardware system that would duplicate the causal powers of the brain to cause consciousness using some chemistry different from neurons." Not much about the necessity of squirting neurotransmitters there. The point I made, and repeat again, is that because we know that brains cause consciousness with specific biological mechanisms, any nonbiological mechanism has to share with brains the causal power to do it. An artificial brain might succeed by using something other than carbon-based chemistry, but just shuffling symbols is not enough, by itself, to guarantee those powers. Once again, he offers no answer to this argument.

He challenges my Chinese Room Argument, but he seriously misrepresents it. The argument is not the circular claim that I do not understand Chinese because I am just a computer, but rather that I don't, as a matter of fact, understand Chinese and could not acquire an understanding by carrying out a computer program. There is nothing circular about that. His chief counterclaim is that the man is only the central processing unit, not the whole computer. But this misses the point of the argument. The reason the man does not understand Chinese is that he does not have any way to get from the symbols, the syntax, to what the symbols mean, the semantics. But if the man cannot get the semantics from the syntax alone, neither can the whole computer. It is, by the way, a misunderstanding on his part to think that I am claiming that a man could actually carry out the billions of steps necessary to carry out a whole program. The point of the example is to illustrate the fact that the symbol manipulations alone, even billions of them, are not constitutive of meaning or thought content, conscious or unconscious. To repeat, the syntax of the implemented program is not semantics.

Concerning other points in his letter: He says that I am wrong to think that he attributes superior thinking to Deep Blue. But here is what he wrote in response to the charge that Deep Blue just does number crunching and not thinking: "One could say that the opposite is the case, that Deep Blue was indeed thinking through the implications of each move and countermove, and that it was Kasparov who did not have the time to think very much during the tournament" (p. 290).

He also says that on his view Moore's Law is only a part of the story. Quite so. In my review I mention other points he makes such as, importantly, nanotechnology.

I cannot recall reading a book in which there is such a huge gulf between the spectacular claims advanced and the weakness of the arguments given in their support. Kurzweil promises us our minds downloaded onto decent hardware, new bodies made of better stuff, evolution without DNA, better sex without the inconvenience of actual partners, computers that convince us that they are conscious, and above all personal immortality. The main theme of my critique is that the existing technological advances that are supposed to provide evidence in support of these predictions, wonderful though they are, offer no support whatever for these spectacular conclusions. In every case the arguments are based on conceptual confusions. Increased computational power by itself is no evidence whatever for consciousness in computers.

3

Organism and Machine: The Flawed Analogy

Michael Denton

11.10.04

The dream of instantiating the properties and characteristics of living organisms in non-living artificial systems is almost as old as human thought. Even in the most primitive of times the magician's model or likeness upon which the rituals of sympathetic magic were enacted was believed to capture some essential quality of the living reality it represented. The magician's likeness, Vaucanson's famous mechanical duck, which was able to eat and drink and waddle convincingly and was one of the wonders of the Paris salons in the eighteenth century, the Golem or artificial man who would protect the Jews of medieval Prague, HAL the life-like computer in the film *2001: A Space Odyssey*, all testify to mankind's eternal fascination with the dream to create another life and to steal

Michael Denton is the Senior Research Fellow in Human Molecular Genetics at the University of Otago in New Zealand, Senior Fellow of Discovery Institute and author of Nature's Destiny *(New York: Free Press, 1998).*

fire from the gods. Ray Kurzweil's book *The Age of Spiritual Machines* represents only one of the latest manifestations of the long-standing dream.

At the outset I think it is important to concede that if living organisms are analogous *in all important respects to artificial mechanical systems* and profoundly analogous to machines—as mechanists since Descartes have always insisted—then in my view there are no serious grounds for doubting the possibility of Kurzweil's "Spiritual Machines." The logic is compelling. Conversely if living things are not machine-like in their basic design—if they differ in certain critical ways from machines as the vitalists have always maintained—then neither artificial life, artificial intelligence nor any of the characteristics of living organisms are likely to be instantiated in non-living mechanical systems.

My approach, therefore, is to question the validity of the machine/organism analogy upon which the whole mechanistic tradition is based. I intend to critique the very presuppositions on which Kurzweil's strong AI project is based, rather than offer detailed analysis of his argument—a task amply provided by other contributors to this volume. Here I am going to argue that there is no convincing evidence that living organisms are strictly analogous to artificial/mechanical objects in the way the mechanist claims and that while certain aspects of life may be captured in artifacts there remains the very real possibility, I would say a near certainty, that elusive, subtle, irreducible "vital" differences exist between the two categories of being the "organic" and the "mechanical." And I would like to suggest that some of the "vital" properties unique to organic systems, which could well include "human intelligence" and perhaps other aspects of what we call "human nature" may never find exact instantiation in artificial manmade systems—a likelihood which would render impossible any sort of "spiritual machine."

The Mechanistic Paradigm

The emergence of the modern mechanistic view of nature and of the idea that organisms are analogous in every essential way to machines—the ultimate source of the thinking of Douglas Hofstadter, Daniel Dennett, Ray Kurzweil and of other supporters of strong AI—coincided roughly with the birth of science in the sixteenth and seventeenth centuries.

One of its first and most influential exponents was the great French philosopher Rene Descartes, for whom the entire material universe was a machine—a gigantic clockwork mechanism. According to this view all of nature—from the movement of the planets to the movements of the heart—worked according to mechanical laws, and all the characteristics of every material object both living and nonliving could be explained in its entirety in terms of the arrangement and movement of its parts. In his own words from his *Treatise on Man*:

> I suppose the body to be nothing but a machine . . . We see clocks, artificial fountains, mills, and other such machines which, although only man made, have the power to move on their own accord in many different ways . . . one may compare the nerves of the machine I am describing with the works of these fountains, its muscles and tendons with the various devices and springs which set them in motion . . . the digestion of food, the beating of the heart and arteries . . . respiration, walking . . . follow from the mere arrangement of the machine's organs every bit as naturally as the movements of a clock or other automaton follow from the arrangements of its counterweights and wheels.

And in his *Principles of Philosophy* he explicitly states:

> I do not recognize any difference between artifacts and natural bodies . . .

Despite occasional set backs ever since, Descartes' biological science has followed by and large along mechanistic lines and nearly all the major advances in knowledge have arisen from its application.

Today almost all professional biologists have adopted the mechanistic/reductionist approach and assume that the basic parts of an organism (like the cogs of a watch) are the primary essential things, that a living organism (like a watch) is no more than the sum of its parts, and that it is the parts that determine the properties of the whole and that (like a watch) a complete description of all the properties of an organism may be had by characterizing its parts in isolation.

The traditional vitalistic alternative has virtually no support. Nowadays few biologists seriously consider the possibility that organic forms (unlike watches) might be natural and necessary parts of the cosmic order—as was believed before the rise of the mechanistic doctrine. Few believe that organisms might be more than the sum of their parts, possessing mysterious vital non-mechanical properties, such as a self-organizing ability or a genuine autonomous intelligence, which are not explicable in terms of a series of mechanical interactions between their parts.

Over and over again the vitalist presumption—that organisms possess special vital powers only manifest by the functioning vital whole—has fallen to the mechanist assault. In the early nineteenth century Wöhler synthesized urea, showing for the first time that organic compounds, previously considered to require living protoplasm for their manufacture, could be assembled artificially outside the cell by non-vital means. Later in the nineteenth century enzymologists showed that the key chemical reactions of the cell could be carried out by cell extracts and did not depend on the intact cell. The triumphant march of mechanism has continued throughout the twentieth century and its application has led to spectacular increases in biological knowledge particularly over the past four decades.

There is no longer any doubt that many biological phenomena are indeed mechanical and that organisms are analogous to machines at least to some degree.

Having achieved so much from the mechanistic approach it is not surprising that the metaphysical assumption of mechanism—that organisms are profoundly analogous to machines in all significant characteristics—is all-pervading and almost unquestioned in modern biology.

Life-like Machines

On top of the undeniable fact that many biological phenomena can be explained in mechanical terms, the credibility of the organism/machine analogy has been reinforced over the past few centuries by our ability to construct increasingly life-like machines.

For most of human history man's tools or machines bore no resemblance to living organisms and gave no hint of any analogy between the living and the artificial. Indeed through most of history, through the long intermittent colds of the Paleolithic, the only machines manufactured by man were primitive bone or wooden sticks and the crudely shaped hand axes—the so-called eoliths or dawn stones. Primitive man was only capable of manufacturing artifacts so crudely shaped that they were hardly distinguishable from natural flakes of rock or pieces of wood and bone. So it is hardly likely that primitive man—although perhaps as intelligent as modern man—would have perceived any analogy between his crudely shaped artifacts and the living beings that surrounded him. Certainly he would never have dreamt of "spiritual machines."

It was not until 10,000 years after the end of the Paleolithic era, following the development of metallurgy, the birth of agriculture and the founding of the first civilizations that humans first manufactured complex artifacts such as ploughs and wheeled vehicles, consisting of several interacting parts. By classical times many artifacts were quite sophisticated, as witness the famous Alexandrian water clock of Ctesibus, Archimedes' screw and the Roman military catapult. Heron of Alexandria wrote several treatises on the construction of lifting machines and presses. The famous device found in a shipwreck off the island of Antiketheria—the Antiketheria computer—

contained a gear train consisting of 31 gears compacted into a small box about the size of a laptop computer. This "computer" was probably a calendrical device for calculating the position of the sun, moon and planets.

Although the technological accomplishments of classical times were quite sophisticated, it was not until the seventeenth century and the beginning of the modern era that technology had advanced to the stage when machines began to take on life-like characteristics.

Ten years before the century opened in 1590, the compound microscope was invented. The telescope was invented in 1608 by the Dutchman Lipershay, shortly afterwards Galileo invented the thermometer, one of his pupils Toricelli the barometer, and in 1654 Guericke invented the air pump. Over the same period clocks and other mechanisms were being vastly improved. At last machines, such as the telescope and microscope with their lens and focusing devices, analogous to that of the vertebrate eye, or hydraulic pumping systems, which were analogous to the action of the heart, began crudely to exhibit some of the characteristics and properties of living things. The fantastic engineering drawings of Leonardo Da Vinci, envisaging futuristic flying and walking machines, lent further support to the machine organism analogy.

The seventeenth and eighteenth centuries also saw the first successful attempts at constructing life-like automata. Vaucanson's duck, for example, which was constructed in about 1730, had over 1,000 moving parts and was able to eat and drink and waddle convincingly. It became one of the wonders of the eighteenth century Parisian salons and represented a forerunner of the robots of today and the androids of science fiction. Such advances raised the obvious possibility that eventually all the characteristics of life including human self-reflective intelligence might find instantiation in mechanical forms.

Since the days of Descartes technology has of course advanced to levels that were simply unimaginable in the seventeenth century. At an ever-accelerating rate one technological advance has followed another. And as machines have continually grown in complexity and sophistication especially since the seventeenth century the gap between living things and machines seems to have continually narrowed. Every few decades machines have seemed to grow more life-like until today there seems hardly a feature of complex machines does not have some analogue in living systems. Like organisms, machines use artificial languages and memory banks for information storage and retrieval. To decode these languages machines, like organisms, use complex translational systems. Modern machinery utilizes elegant control systems regulating the assembly of parts and components, error fail-safe devices and proofreading systems are utilized for quality control, assembly processes utilize the principle of prefabrication. All these phenomena have their parallel in living systems. In fact, so deep and so persuasive is the analogy that much of the terminology we use to describe the world of the cell is borrowed from the world of late twentieth century technology.

From stone axe-head to modern technology mankind has journeyed far, very far since the long colds of the Paleolithic dawn. And given the increasing "life-likeness" of many modern artifacts it seems likely, or so the mechanist would have us believe, that eventually all the phenomena of life will be instantiated in mechanical forms. Surely the day of *Spiritual Machines* can hardly be that far away?

Organic Form: Vital Characteristics

Yet despite the obvious successes of mechanistic thinking in biology and the fact that many biological phenomena can be reduced to mechanical explanations, and despite the fact that machines have grown ever more life-like as technology has advanced—it remains an undeniable fact that living things possess abilities that are still without any significant analogue in any machine which has yet been constructed. These abilities have been seen since classical times as indicative of a fundamental division between the vital and mechanical modes of being.

To begin with, every living system replicates itself, yet no machine possesses this capacity even to the slightest degree. Nor has any machine—even the most advanced envisaged by nanotechnologists—been conceived of that could carry out such a stupendous act. Yet every second countless trillions of living systems from bacterial cells to elephants replicate themselves on the surface of our planet. And since life's origin, endless life forms have effortlessly copied themselves on unimaginable numbers of occasions.

Living things possess the ability to change themselves from one form into another. For example, during development the descendants of the egg cell transform themselves from undifferentiated unspecialized cells into wandering amoebic cells, thin plate-like blood cells containing the oxygen-transporting molecule hemoglobin, neurons—cells sending out thousands of long tentacles like miniature medusae some hundred thousand times longer than the main body of the cell.

The ability of living things to replicate themselves and change their form and structure are truly remarkable abilities. To grasp just how fantastic they are and just how far they transcend anything in the realm of the mechanical, imagine our artifacts endowed with the ability to copy themselves and—to borrow a term from science fiction—"morph" themselves into different forms. Imagine televisions and computers that duplicate themselves effortlessly and which can also "morph" themselves into quite different types of machines—a television into a microwave cooker, or a computer into helicopter. We are so familiar with the capabilities of life that we take them for granted, failing to see their truly extraordinary character.

Even the less spectacular self re-organizing and self regenerating capacities of living things—some of which have been a source of wonderment since classical times—should leave the observer awestruck. Phenomena such as the growth of a complete tree from a small twig, the regeneration of the limb of a newt, the growth of a complete polyp, or a complex protozoan from tiny fragments of the intact animal are again phenomena without analogue in the realm of mechanism. To grasp something of the transcending nature of this remarkable phenomenon, imagine a jumbo jet, a computer, or indeed any machine ever conceived from the fantastic star ships of science fiction to the equally fantastic speculations of nanotechnology, being chopped up randomly into small fragments. Then imagine every one of the fragments so produced (no two fragments will ever be the same) assembling itself into a perfect but miniaturized copy of the machine from which it originated—a tiny toy-sized jumbo jet from a random section of the wing—and you have some conception of the self regenerating capabilities of certain microorganisms such as the ciliate *Stentor*. It is an achievement of transcending brilliance, which goes beyond the wildest dreams of mechanism.

And it is not just the self-replicative, "morphing" or self-regenerating capacity which has not been instantiated in mechanical systems. Even the far less ambitious end of component self-assembly has not been achieved to any degree. This facility is utilized by every living cell on earth and is exhibited in processes as diverse as protein folding, the assembly of viral capsules and the assembly of cell organelles such as the ribosome. In these processes tens or hundreds of unique and individually complex elements combine together, directed entirely by their own intrinsic properties without any external intelligent guidance or control is an achievement without any analogue in modern technology in spacecraft, in computers, or even in the most outrageous speculations of nanotechnologists. Imagine a set of roughly hewn but quite unfinished components of a clock—several dozen ill-fitting cogs, wheels, axles, springs and a misshapen clock face completely incapable of fitting together to make a clock in their current "primary" unfinished state, so out of alignment and so imperfectly hewn, so different in form from their final state and purpose that the end they are intended to form could never be inferred. Now imagine the set to be animated by some magic force and beginning to come together piece by piece, this cog and that cog, this wheel and this axle. Imagine that as they interact together each changes or more properly reshapes its neighbor so that they both come to fit perfectly together; and so through a series of such mutual self-forming activities the whole animated set of components is transformed, again, as if by magic, into the form of a functioning clock. It is as if the original parts "knew" the end for which they were intended and had the ability to fashion themselves towards that end, as if an invisible craftsman were fashioning the parts to their ordained end. It is not hard to see how Aristotle came to postulate an entelechy or soul as the immanent directive force organizing matter to its ordained end.

Such an animated self-assembly process, like the formation of a whole protozoan from a tiny fragment of the cell, is another vital capacity of transcending brilliance absolutely unparalleled in any mechanical system.

Finally I think it would be acknowledged by even ardent advocates of strong AI like Kurzweil, Dennett and Hofstadter that *no machine has been built to date which exhibits consciousness and can equal the thinking ability of humans.* Kurzweil himself concedes this much in his book. As he confesses: "Machines today are still a million times simpler than the human brain. Their complexity and subtlety is comparable to that of insects." Of course Kurzweil believes, along with the other advocates of strong AI that sometime in the next century computers capable of carrying out 20 million billion calculations per second (the capacity of the human brain) will be achieved and indeed surpassed. And in keeping with the mechanistic assumption that organic systems are essentially the same as machines then of course such machines will equal or surpass the intelligence of man. My prediction would be that such machines will be wonderfully fast calculators but will still not possess the unique characteristics of the human brain, the ability for conscious rational self-reflection. In his book *Gödel, Escher, Bach,* Hofstader—himself a believer in the possibility of strong AI—makes the point explicitly that the entire AI project is dependent on the mechanistic or reductionist faith.

Although the mechanistic faith in the possibility of strong AI still runs strong among researchers in this field, Kurzweil being no exception, there is no doubt that *no one has manufactured anything that exhibits intelligence remotely resembling that of man.*

It is clear then that living systems *do* exhibit certain very obvious characteristics including intelligence, the capacity for self-replication, self-assembly, self-reorganization and for morphological transformations which are without analogy in any human contrivance. Moreover, these are precisely the characteristics which have been viewed since classical times as the essential defining characteristics of the vital or organic realm.

Organic Form: A Unique Holistic Order

In addition to possessing the unique abilities discussed above, it is also evident that the basic design of organic systems from individual macromolecules to embryos and brains exhibits a unique order which is without analogy in the mechanical realm. This unique order involves a reciprocal formative influence of all the parts of an organic whole on each other and on the whole in which they function.

The philosopher Immanuel Kant clearly recognized that this "reciprocal formative influence of the parts on each other" is a unique characteristic of organic form. In his famous analysis of organic form in *Critique of Teleological Judgment* he argues that an organism is a being in which "the parts . . . combine themselves into the unity of a whole by being reciprocally cause and effect of their form. . . . (and this unity) may reciprocally determine in its turn the form and combination of all the parts." He continues:

> . . . In such a natural product as this every part is thought as owing its presence to the agency of all the remaining parts and also as existing for the sake of the others and of the whole . . . the part must be an organ producing the other parts, each consequently reciprocally producing the others.

He then contrasts the reciprocal formative influence of the parts of organisms with the non-formative relationships between parts in mechanical wholes:

In a watch, one part is the instrument by which the movement of the others is affected, but one wheel is not the efficient cause of the production of the other. One part is certainly present for the sake of another, but it does not owe its presence to the agency of that other. For this reason also the producing cause of the watch and its form is not contained in the nature of this material. Hence one wheel in the watch does not produce the other and still less does one watch produce other watches by utilizing or organizing foreign material. Hence it does not of itself replace parts of which it has been deprived.

Kant concludes with an insightful definition of organisms as beings "in which every part is both means and end, cause and effect." In his view such " an organization has nothing analogous to any causality known to us." He refers to it as an "impenetrable property" of life.

Perhaps no organic forms illustrate this unique order more clearly than the simplest of all organic forms, the vital building blocks of all life—the proteins. Proteins are the very stuff of life. All the vital chemical functions of every cell on earth are all in the last analysis dependent on the activities of these tiny biological systems, the smallest and simplest of all the known systems of organic nature. Proteins are also the basic building blocks of life for it is largely by the association of different protein species that all the forms and structures of living things are generated.

It is immediately obvious even to someone without any previous experience in molecular biology or without any scientific training that the arrangement of the atoms in a protein is unlike any ordinary machine or any machine conceived. Indeed the protein is unlike any object of common experience. One superficial observation is the apparent illogic of the design and the lack of any obvious modularity or regularity. The sheer chaos of the arrangement of the atoms conveys an almost eerie other-worldly non-mechanical impression.

Interestingly a similar feeling of the strangeness and chaos of the arrangement of atoms in a protein struck the researchers at Cambridge University after the molecular structure of the first protein, myoglobin, had been determined in 1957 (using the technique of X-ray crystallography). Something of their initial feelings are apparent in Kendrew's comments at the time (reported by M. Perutz in the *European Journal of Biochemistry* 8 (1969): 455-466:

> Perhaps the most remarkable features of the molecule are its complexity and its lack of symmetry. The arrangement seems to be almost totally lacking in the kind of regularities which one instinctively anticipates, and it is more complicated than had been predicted by any theory of protein structure.

In the late fifties, as the first three-dimensional structures of proteins were worked out, it was first assumed—in conformity with mechanistic thinking—that each amino acid made an individual and independent contribution to the 3D form of the protein. This simplifying assumption followed from the concept of proteins as "molecular machines" in the literal sense. This implied that their design should be like that of any machine, essentially modular, built up from a combination of independent parts each of which made some unique definable contribution to the whole.

It soon became apparent, however, that the design of proteins was far more complex than scientists first assumed. In fact, the contribution of each individual amino acid to the tertiary configuration of a protein was not straightforward but was influenced by subtle interactions with many of the other amino acids in the molecule. After thirty years of intensive study it is now understood that the spatial conformation adopted by each segment of the amino acid chain of a protein is specified by a complex web of electronic or electro-chemical interactions, including hydrogen bonds and hydrophobic forces, which ultimately involve directly via short range or indirectly via long range interaction virtually every other section of the amino acid

chain in the molecule. It might be claimed with only slight exag-
geration that the position of each one of the thousands of atoms is
influenced by all the other atoms in the molecule and that each atom
contributes via immensely complex co-operative interactions with
all the other atoms in the protein, something to the overall shape and
function of the whole molecule.

The universal occurrence in proteins of various structural motifs
such as alpha helices and beta sheets conveys the impression that
they represent independent or relatively independent components or
separable modules. On the contrary, the stability and form of such
motifs is determined by a combination of short-range interactions
within the motif and the microchemical environment which it occu-
pies within the molecule which is in turn generated by the global
web of interactions between all the constituent atoms of the protein.
This is evidenced by the fact that the same amino acid sequence often adopts quite different secondary structural conformations in different proteins. The form and properties of each component or part
of a protein or group of atoms—whether it is a major structural motif
or a small twist in the amino acid chain—is dependent on its chemical and physical context or environment within the protein. This context is itself generated by the summation of all the chemical and physical forces, which make up the whole undivided protein itself.

There is no doubt then that proteins are very much less modular
than machines, which are built up from a set of relatively independent modules or compartments. Remove the cog from a watch and it
still remains a cog, remove the wheel from a car and it remains a
wheel. Remove a fragment of a protein and its form disassembles.
What a protein represents is an object in which all the "parts" are in
a reciprocal formative relationship with each other and with the whole.
The parts of the final whole are shaped and finished by reciprocal
interaction with each other.

In the four decades since researchers determined the 3D atomic
configuration of the first protein (myoglobin), we have learned much
about the molecular biology of these remarkable molecules. Although
still widely described in the biological literature as molecular ma-

chines, proteins transcend mechanism in their complexity, in the intense functional integration and interdependence of all their components, in the holistic way that the form and function of each part is determined by the whole and vice versa and in the natural formative process by which the amino acid chain is folded into the native function. In these ways, they resemble no structure or object constructed or conceived by man.

What is true of proteins is also true of the other class of large macromolecules in the cell—the RNA molecules. These too fold into complex three-dimensional forms in which all the parts in the final form are shaped by similar reciprocal formative interactions with the other parts of the molecule. Again the distinctive shapes and forms of the constituent parts only exist in the whole. When removed from the whole, they take on a different form and set of properties or disassemble into a random chain.

The next level of biological complexity above the individual protein and RNA molecules are the multiprotein complexes that make up most of the cell's mass and carry out all the critical synthetic, metabolic and regulatory activities on which the life of the cell depends. These include complexes such as the ribosome (the protein-synthesizing apparatus), which contains more than 55 different protein molecules and three RNA molecules, the transcriptional apparatus which makes the mRNA copy of the gene, which again consists of more than 20 individual proteins, and a variety of higher order structural complexes including the cytoskeleton system, which consists of a highly complex integrated set of microtubules, microfilaments and intermediate fibers.

It is true of virtually all these multimolecular assemblies that the parts in their final form—just like the parts of a protein—have a reciprocal formative influence on each other. So again, in a very real sense the parts do not exist outside the whole. The ribosome illustrates. The assembly of the ribosome takes place in a number of stages involving the stepwise association of the 55 proteins with each other and with the three RNA molecules. As the assembly progresses the growing RNA-protein complex undergoes conformational changes, some relatively minor and some major, until the final mature functional form of the ribosome is generated. The process is cooperative, with all parts of the growing particle having a formative influence on all other parts either directly, because the "parts" are adjacent, or indirectly through global influences.

The next readily recognizable level we reach as we ascend the organic hierarchy is the living cell. Again the parts of a cell, like those of a protein or a ribosome, are existentially dependent on their being "parts" of the greater whole—the totality of dynamic activities which make up the life of the whole cell. Take them out of the cell and eventually all will die and disintegrate, even if some of their activities may persist long enough for in vitro experimental analyses. Every process, every structure exists only as a result of the dynamic interaction of the whole cell in which they function.

The reciprocal formative relationship between parts and between parts and whole which is observed in proteins, multimolecular systems and cells is also characteristic of all higher order organic structures—including organs like the brain and whole developing embryos. Again, in all cases the parts are existentially dependent on being part of the whole in which they function.

Organic Form: The Failure of Reduction

It is primarily because of the unique holistic order of organic form, in which parts are existentially dependent on their being in the whole and have no real existence outside the whole, that the mechanist goal of providing a complete explanation of the behavior and properties of organic wholes from a complete characterization of the parts in isolation is very far from being achieved.

From knowledge of the genes of an organism it is impossible to predict the encoded organic forms. Neither the properties nor structure of individual proteins nor those of any higher order forms—such as ribosomes and whole cells—can be inferred even from the most exhaustive analysis of the genes and their primary products, linear sequences of amino acids. In a very real sense organic forms from proteins to the human mind are not specified in the genes but rather arise out of reciprocal self-formative interactions among gene products and represent genuinely emergent realities which are ever-present—at least in potential—in the order of nature.

And it is precisely because of the impossibility of prediction and design of organic form from below, that engineering new biological forms is fantastically difficult. In these days of genetic engineering we hear so much about "transforming life" and "re-designing organisms" that it comes as something of a surprise to learn that not even one completely new functional protein containing only 100 subunits (amino acids) has been designed successfully, let alone something as complex as a new type of virus or a cellular organelle like a ribosome, objects containing on the order of 10,000 subunits (amino acids and nucleotides).

The total failure of reductionism in the realm of the organic and the total failure to engineer new forms, contrasts markedly with the situation in the realm of the mechanical. In the case of machines from jumbo jets to typewriters, the properties and behavior of the whole can be predicted entirely and with great accuracy from below, that is, from an exhaustive characterization of their parts in isolation. It is because the parts of machines do not undergo complex reciprocal self-formative interactions but have essentially the same properties and form in isolation as they do in the mechanical whole that makes their design possible. Machines are no more than the sum of their parts in isolation. And this is why we have no trouble assembling complex artifacts like space shuttles or jet airliners that contain more than a million unique components. Yet no artifact has ever been built, even one consisting of only 100 components (the same number of components as in a simple protein), which exhibits a reciprocal self-formative relationship between its parts. This unique property, as we have seen above, is the hallmark of organic design.

Organic design is essentially a top-down reality. As we have seen, organic forms are essentially nonmodular wholes and their order is intrinsic to, and only manifest in, the functioning whole. Success in engineering new organic forms from proteins up to organisms will therefore require a completely novel approach, a sort of designing from "the top down." Because the parts of organic wholes only exist in the whole, organic wholes cannot be specified bit by bit and built up from a set of relatively independent modules; consequently the entire undivided unity must be specified together *in toto*.

If proteins and other higher order organic forms *had* been built up mechanically out of modules, each of which had the same form in an isolated state that it has in the final form—rather like the parts of a machine, the cogs of a watch or pieces in a child's erector set such as *Legos*—then the problem of predicting organic form would have been solved years ago. By now the world would be full with all manner of new "artificial life forms"

The Organic and the Mechanical: Two Distinct Categories of Being

From the evidence discussed above it is clear that the machine/organism analogy is only superficial. Although organisms do exhibit mechanical or machine-like properties they also possess properties which no machine exhibits even to a minor degree. In addition, the design of life exhibits a "holistic" order which is without parallel in the realm of the mechanical and which cannot be reduced to or explained in mechanical terms. Consequently, no new organic form has been engineered to date. There is self-evidently a gulf between the organic and the mechanical, which has not been bridged.

The picture of organic form that has emerged from recent advances in biology is surprisingly consistent with the pre-mechanistic holistic/vitalistic model, which was first clearly formulated by the Greeks and specifically by Aristotle. According to their view, each organic whole or form was believed to be a unique integral whole or indivisible unity. This whole—in effect a self-sufficing soul or entelechy—regulated, organized and determined the form, properties and behavior of all of its component parts. Taken to its logical conclusion this model implied that only wholes have any real autonomous existence and that the parts of wholes have no independent existence or meaning outside the context of the whole.

Aristotle expressed this concept in *The Parts of Animals* :

> When any one of the parts or structures, be that which
> it may, is under discussion, it must not be supposed
> that it is its material composition to which attention is
> being directed or which is the object of the discus-
> sion, but the relation of such part to the total form.
> Similarly, the true object of architecture is not bricks,
> mortar, or timber, but the house; and so the principal
> object of natural philosophy is not the material ele-
> ments, but their composition, and the totality of the
> form, *independently of which they have no existence*.

For the Greeks all natural objects including organic forms were
an integral part of the world order, or Cosmos. Each organic whole
or form was therefore an eternal unchangeable part of the basic fab-
ric of the universe. Each was "a potential awaiting actualization"
"animated by an immanent finality," in the words of Genevieve Rodis
-Lewis. Its design, the purposeful arrangement of parts to wholes,
was internal to the organism itself—an outcome of the "inner developmental force which impelled it towards the realization of its form,"
[Rodin-Lewis G (1978). "Limitations of the Mechanical Model in
the Cartesian Conception of the Organism," in *Descartes*, ed. M.
Hooker (Baltimore: John Hopkins University Press, Pp. 152- 170.)]
Or in the words of the *Cambridge Companion to Aristotle* the organic form is "an internal structural principle striving to actualize
itself as the fully mature individual." (*Cambridge Companion to
Aristotle*, Ed. J Barnes, 1995). As Jonathon Lear puts it in his *Aristotle:
The Desire to Understand*: Natural organisms "are foci of reality and
self-determination . . . possessing...an inner life of their own organisms occupying a fundamental ontological position: they are among
the basic things that are."

This is not the place for a detailed exposition or defense of the
Greek conception of the organism. Suffice to say that for the Greeks
organisms were a totally different type of being to that of lifeless

artifacts. If we leave out the soul we are left with a holistic model of the organic world, which is very close to that revealed by modern biological science.

In my view the fact that organisms and machines belong to different categories of being is beyond dispute, even if the final nature of this fundamental difference is not yet clearly defined. And surely the existence of this "vital" difference raises profound doubts as to the validity of Kurzweil's claim. For, if we are incapable of instantiating in mechanical systems any of the "lesser" vital characteristics of organisms such as self-replication, "morphing," self-regeneration, self-assembly and the holistic order of biological design, why should we believe that the most extraordinary of all the "vital characteristics"—the human capacity for conscious self-reflection—will ever be instantiated in a human artifact?

And it is not just Kurzweil's claims that are in doubt. If the traditional vitalist position is true, and organic forms are integral parts of the cosmic order, each being in essence an indivisible self-sufficing unity possessing properties and characteristics beyond those possessed by any machine, properties which might include for example intelligent self-organizing capabilities, then *all nature becomes re-enchanted*. The whole mechanistic framework of modern biology would have to be abandoned, along with many key doctrines, including the central dogma of genetic determinism, the concept of the "passivity and impotence" of the phenotype and the spontaneity of mutation. Moreover all theories of the origin and evolution of life and biological information would have to be re-formulated in conformity with vitalistic principles and all explanations based on the mechanistic concept of organisms as fundamentally lifeless contingent combinations of parts—including contemporary Darwinism—would have to be revised. Even certain traditional design arguments such as Paley's would have to be reconsidered, since they presume that organisms are analogous to artifacts, being in essence contingent and unnecessary, and thus, like human artifacts, require an artificer or craftsmen for their assembly.

4

Kurzweil's Impoverished Spirituality

William A. Dembski

12.2.04

Tender-Minded Materialism

The question of whether humans are machines has been vigorously debated over the last two hundred years. The French materialists of the Enlightenment like Pierre Cabanis, Julien La Mettrie, and Baron d'Holbach affirmed that humans are machines (La Mettrie even wrote a book titled *Man the Machine*). Likewise contemporary materialists like Marvin Minsky, Daniel Dennett, and Patricia Churchland see the motions and modifications of matter as sufficient to account for human mentality. For all its faults, materialism is a predictable philosophy. If matter is all there is, then mind must, in some fashion, reduce to matter. Whereas the

William Dembski is Associate Research Professor of the Conceptual Foundations of Science at Baylor University, Senior Fellow of Discovery Institute and author of The Design Inference *(Cambridge: Cambridge U. Press, 1998).*

Enlightenment philosophes might have thought of humans in terms of gear mechanisms and fluid flows, contemporary materialists think of humans in terms of neurological systems and computational devices. The idiom has been updated, but the underlying impulse to reduce mind to matter remains unchanged.

If predictability is materialism's main virtue, then hollowness is its main fault. Humans have aspirations. We long for freedom, immortality, and the beatific vision. We are restless until we find our rest in God. The problem for the materialist, however, is that these aspirations cannot be redeemed in the coin of matter. Our aspirations are, after all, spiritual (etymology confirms this point—"aspiration" and "spiritual" are cognates). We need to transcend ourselves to find ourselves. Now the motions and modifications of matter offer no opportunity for transcending ourselves. Materialists in times past admitted as much. Freud saw belief in God as wish-fulfillment. Marx saw religion as an opiate. Nietzsche saw Christianity as a pathetic excuse for mediocrity. Each regarded the hope for transcendence as a delusion.

This hope, however, is not easily excised from the human heart. Even the most hardened materialist shudders at Bertrand Russell's vision of human destiny: "Man is the product of causes which had no prevision of the end they were achieving" and which predestine him "to extinction in the vast death of the solar system." The human heart longs for more. And in an age when having it all has become *de rigueur*, enjoying the benefits of religion without its ontological burdens is now within reach. The erstwhile impossible marriage between materialism and spirituality is now routinely consummated. The vision of C.S. Lewis' devilish character Screwtape—the "materialist magician" who combines the skepticism of the materialist with the cosmic consciousness of the mystic—is here at last.

Within the tough-minded materialism of the past, human aspirations, whatever else they might be, were strictly finite and terminated with the death of the individual. The tough-minded materialism of the past was strong, stark, and courageous. It embraced the void, and disdained any impulse to pie in the sky. Not so the tender-minded materialism of our age. Though firmly committed to materialism, it is just as firmly committed to not missing out on any benefits ascribed to religious experience. A spiritual materialism is now possible, and with it comes the view that we are spiritual machines. The juxtaposition of spirit and mechanism, which previously would have been regarded as an oxymoron, is now said to constitute a profound insight.

As evidence for this move from tough- to tender-minded materialism, consider Ray Kurzweil's recently published *The Age of Spiritual Machines: When Computers Exceed Human Intelligence*. Kurzweil is a leader in artificial intelligence, and specifically in the field of voice-recognition software. Ten years ago Kurzweil published the more modestly titled *The Age of Intelligent Machines* (MIT, 1990). There he gave the usual strong artificial intelligence position about machine and human intelligence being functionally equivalent. In *The Age of Spiritual Machines*, however, Kurzweil's aim is no longer to show that machines are merely capable of human capacities. Rather, his aim is to show that machines are capable of vastly outstripping human capacities and will do so within the next thirty years.

According to *The Age of Spiritual Machines*, machine intelligence is the next great step in the evolution of intelligence. That the highest form of intelligence happens for now to be embodied in human beings is simply an accident of natural history. Human beings need to be transcended, though not by going beyond matter, but by reinstantiating themselves in more efficient forms of matter, to wit, the computer. Kurzweil claims that in the next thirty or so years we shall be able to scan our brains, upload them onto a computer, and thereafter continue our lives as virtual persons running as programs on machines. Since the storage and processing capacities of these virtual persons will far exceed that of the human brain, they will quickly take the lead in all aspects of society. Those humans who refuse to upload themselves will be left in the dust, becoming "pets," as Kurzweil puts it, of the newly evolved computer intelligences. What's more, these computer intelligences will be conditionally immortal, depending for their continued existence only on the ability of hardware to run the relevant software.

Although Kurzweil is at pains to shock his readers with the imminence of a computer takeover, he is hardly alone in seeking immortality through computation. Frank Tipler's *The Physics of Immortality* (Doubleday, 1994) is devoted entirely to this topic. Freeman Dyson has pondered it as well. Alan Turing, one of the founders of modern computation, was fascinated with how the distinction between software and hardware illuminated immortality. Turing's friend Christopher Morcom had died when they were teenagers. If Morcom's continued existence depended on his particular embodiment, then he was gone for good. But if he could be instantiated as a computer program (software), Morcom's particular embodiment (hardware) would be largely irrelevant. Identifying personal identity with computer software thus ensured that people were immortal since even though hardware could be destroyed, software resided in a realm of mathematical abstraction and was thus immune to destruction.

Humans as Spiritual Machines

A strong case can be made that humans are not machines—period. I shall make that case later in this essay. Assuming that I am right and that humans are not machines, it follows that humans are not spiritual machines. Even so, the question what it would mean for a machine to be spiritual is interesting in its own right. My immediate aim, therefore, is not to refute the claim that humans are spiritual machines, but to show what modifying "machine" with the adjective "spiritual" entails. I shall argue that attributing spirituality to machines entails an impoverished form of spirituality. It's rather like talking about "free prisoners." Whatever else freedom might mean here, it doesn't mean freedom to leave the prison.

To see what modifying "machine" with the adjective "spiritual" entails, let us start by examining what we mean by a machine. Normally by a machine we mean an integrated system of parts that function together to accomplish some purpose. To avoid the troubled waters of teleology, let us bracket the question of purpose. In that case we can define a machine as any integrated system of parts whose motions and modifications entirely characterize the system. Implicit in this definition is that all the parts are physical. Consequently a machine is fully determined by the constitution, dynamics, and interrelationships of its physical parts.

This definition is very general. It incorporates artifacts as well as organisms (humans being a case in point). Because the nineteenth century Romanticism that separates organisms from machines is still with us, many people shy away from calling organisms machines. But organisms are as much integrated systems of physical parts as are human artifacts. Perhaps "integrated physical systems" would be more precise, but "machines" stresses the strict absence of extra-material factors from such systems, and it is that absence which is the point of controversy.

With this definition of machines in hand, let us now consider what it means to ascribe spirituality to machines. Because machines are integrated systems of parts, they are subject to what I call the *replacement principle*. What this means is that physically indistinguishable parts of a machine can be exchanged without altering the machine. At the subatomic level, particles in the same quantum state can be exchanged without altering the subatomic system. At the biochemical level, polynucleotides with the same length and sequence specificity can be exchanged without altering the biochemical system. At the organismal level, identical organs can be exchanged without altering the biological system. At the level of human contrivances, identical components can be exchanged without altering the contrivance.

The replacement principle is relevant to this discussion because it implies that machines have no substantive history. According to Hilaire Belloc, "To comprehend the history of a thing is to unlock the mysteries of its present, and more, to disclose the profundities of its future." But a machine, properly speaking, has no history. Its history is a superfluous rider—an addendum that could easily have been different without altering the machine. If something is solely a machine, then according to the replacement principle it and a replica are identical. Forgeries of the present become masterpieces of the past if the forgeries are good enough. This may not be a problem for art dealers, but it does become a problem when the machines in question are ourselves (cf. matter compilers that à la *Star Trek* could assemble and diassemble us atom by atom).

For a machine, all that it is, is what it is at this moment. We typically think of our memories as either remembered or forgotten, and if forgotten then having the possibility of recovery. But machines do not properly speaking remember or forget (remembering and forgetting being substantive relations between a person and a person's actual past). Machines access or fail to access items in storage. What's more, if they fail to access an item, it's either because the retrieval mechanism failed or because the item was erased. Consequently, items that represent past occurrences but were later erased are, as far as the machine is concerned, just as though they never happened. Mutatis mutandis, items that represent counterfactual occurrences (i.e., things that never happened) but which are accessible can be, as far as the machine is concerned, just as though they did happen.

The causal history leading up to a machine is strictly an accidental feature of it. Consequently, any dispositional properties we ascribe to a machine (e.g., goodness, morality, virtue, and yes, even spirituality) properly pertain only to its current state and future potentialities, and can be detached from its past. In particular, any defect in a machine relates only to its current state and future potentialities. Moreover, the correction of any defect properly belongs to technology. A machine that was a mass-murderer yesterday may become an angel of mercy today provided we can find a suitable readjustment of its parts. Having come to view ourselves as machines, it is no accident that our society looks for salvation in technologies like behavior modification, psychotropic drugs, cognitive reprogramming, and genetic engineering.

The problem with machines is that they are incapable of sustaining what philosophers call *substantial forms*. A substantial form is a principle of unity that holds a thing together and maintains its identity over time. Machines lack substantial forms. A machine, though having a past, might just as well not have had a past. A machine, though configured in one way, could just as well be reconfigured in other ways. A machine's defects can be corrected and its virtues improved through technology. Alternatively, new defects can be introduced and old virtues removed through technology. What a machine is now and what it might end up in the future are entirely open-ended and discontinuous. Despite the buffeting of history, substantial forms perdure through time. Machines, on the other hand, are the subject of endless tinkering and need bear no semblance to past incarnations.

In this light consider the various possible meanings of "spiritual" in combination with "machine." Since a machine is entirely characterized in terms of the constitution, dynamics, and interrelationships of its physical parts, "spiritual" cannot refer to some nonphysical aspect of the machine. Let's therefore restrict "spiritual" to some physical aspect of a machine. What, then, might it refer to? Often when we think of someone as spiritual, we think of that person as exhibiting some moral virtue like self-sacrifice, altruism, or courage. But we only attribute such virtues on the basis of past actions. Yet past actions belong to history, and history is what machines don't have, except accidentally.

Consider, for instance, a possible-worlds scenario featuring an ax murderer who just prior to his death has a cerebral accident that turns his brain state into that of Mother Teresa's at her most charitable. The ax murderer now has the brain state of a saint but the past of a sinner. Assuming the ax murderer is a machine, is he now a spiritual machine? Suppose Mother Teresa has a cerebral accident just prior to her death that turns her brain state into that of the ax murderer's at his most barbaric. Mother Teresa now has the brain state of a sinner but the past of a saint. Assuming Mother Teresa is a machine, is she no longer a spiritual machine?

Such counterfactuals indicate the futility of attributing spirituality to machines on the basis of past actions. Machines that have functioned badly in the past are not sinners and therefore unspiritual. Machines that have functioned well in the past are not saints and therefore spiritual. Machines that have functioned badly in the past need to be fixed. Machines that have functioned well in the past need to be kept in good working order so that they continue to function well. Once a machine has been fixed, it doesn't matter how badly it functioned in the past. On the other hand, once a machine goes haywire, it doesn't matter how well it functioned in the past.

Attributing spirituality to machines on the basis of future actions is equally problematic. Clearly, we have access to a machine's future only through its present. Given its present constitution, can we predict what the machine will do in the future? The best we can do is specify certain behavioral propensities. But even with the best propensities, machines break and malfunction. It is impossible to predict the full range of stresses that a machine may encounter and that may cause it to break or malfunction. Consequently it is impossible to tell whether a machine that gives all appearances of functioning one way will continue to function that way. For every machine in a given state there are circumstances sure to lead to its undoing. Calling a machine "spiritual" in reference to its future can therefore only refer to certain propensities of the machine to function in certain ways. But spirituality of this sort is better left to a bookmaker than to a priest or guru.

Since the future of a machine is accessed through its present, it follows that attributing spirituality to machines properly refers to some present physical aspect of the machine. But what aspect might this be? What about the constitution, dynamics, and interrelationships of a machine's parts renders it spiritual? What emergent property of a system of physical parts corresponds to spirituality? Suppose humans are machines. Does an ecstatic religious experience, an LSD drug trip, a Maslow peak experience, or a period of silence, prayer, and meditation count as a spiritual experience? I suppose if we are playing a Wittgensteinian language game, this usage is okay. But however we choose to classify these experiences, it remains that machine spirituality is the spirituality of immediate experience. This is of course consistent with much of contemporary spirituality, which places a premium on religious experience and neglects such traditional aspects of spirituality as revelation, tradition, virtue, morality, and above all communion with a nonphysical God who transcends our physical being.

Machine spirituality neglects much that has traditionally been classified under spirituality. From this alone it would follow that machine spirituality is an impoverished form of spirituality. But the problem is worse. Machine spirituality fails on its own terms as a phenomenology of religious experience. The spiritual experience of a machine is necessarily poorer than the spiritual experience of a being that communes with God. The entire emphasis of Judeo-Christian spirituality is on communion with a free personal transcendent God (cf. Diogenes Allen's *Spiritual Theology*, Cowley Publications, 1997). Moreover, communion with God always presupposes a free act by God to commune with us. Freedom here means that God can refuse to commune with us (to, as the Scriptures say, "hide his face"). Thus, within traditional spirituality we are aware of God's presence because God has freely chosen to make his presence known to us. Truly spiritual persons—or *saints* as they are called—experience a constant, habitual awareness of God's presence.

But how can a machine be aware of God's presence? Recall that machines are entirely defined by the constitution, dynamics, and interrelationships among their physical parts. It follows that God cannot make his presence known to a machine by acting upon it and thereby changing its state. Indeed, the moment God acts upon a machine to change its state, it no longer properly is a machine, for an aspect of the machine now transcends its physical constituents. It follows that awareness of God's presence by a machine must be independent of any action by God to change the state of the machine. How then does the machine come to awareness of God's presence? The awareness must be self-induced. Machine spirituality is the spirituality of self-realization, not the spirituality of an active God who freely gives himself in self-revelation and thereby transforms the beings with which he is in communion. For Kurzweil to modify "machine" with the adjective "spiritual" therefore entails an impoverished view of spirituality.

Accounting for Intelligent Agency

The question remains whether humans are machines (with or without the adjective "spiritual" tacked in front). To answer this question, we need first to examine how materialism understands human agency and, more generally, intelligent agency. Although the materialist literature that attempts to account for human agency is vast, the materialist's options are in fact quite limited. The materialist world is not a mind-first world. Intelligent agency is therefore in no sense prior to or independent of the material world. Intelligent agency is a derivative mode of causation that depends on underlying natural—and therefore unintelligent—causes. Human agency in particular supervenes on underlying natural processes, which in turn usually are identified with brain function.

How well have natural processes been able to account for intelligent agency? Cognitive scientists have achieved nothing like a full reduction. The French Enlightenment thinker Pierre Cabanis remarked: "Les nerfs—voilà tout l'homme" (the nerves—that's all there is to man). A full reduction of intelligent agency to natural causes would give a complete account of human behavior, intention, and emotion in terms of neural processes. Nothing like this has been achieved. No doubt, neural processes are correlated with behavior, intention, and emotion. But correlation is not causation.

Anger presumably is correlated with certain localized brain excitations. But localized brain excitations hardly explain anger any better than overt behaviors associated with anger, like shouting obscenities. Localized brain excitations may be reliably correlated with anger, but what accounts for one person interpreting a comment as an insult and experiencing anger, and another person interpreting that same comment as a joke and experiencing laughter? A full materialist account of mind needs to understand localized brain excitations in terms of other localized brain excitations. Instead we find localized brain excitations (representing, say, anger) having to be explained in terms of semantic contents (representing, say, insults). But this mixture of brain excitations and semantic contents hardly constitutes a materialist account of mind or intelligent agency.

Lacking a full reduction of intelligent agency to natural processes, cognitive scientists speak of intelligent agency as *supervening* on natural processes. Supervenience is a hierarchical relationship between higher order processes (in this case intelligent agency) and lower order processes (in this case natural processes). What supervenience says is that the relationship between the higher and lower order processes is a one-way street, with the lower determining the higher. To say, for instance, that intelligent agency supervenes on neurophysiology is to say that once all the facts about neurophysiology are in place, all the facts about intelligent agency are determined as well. Supervenience makes no pretense at reductive analysis. It simply asserts that the lower level determines the higher level—how it does it, we don't know.

Certainly, if we knew that materialism were correct, then supervenience would follow. But materialism itself is at issue. Neuroscience, for instance, is nowhere near underwriting materialism, and that despite its strident rhetoric. Hardcore neuroscientists, for instance, refer disparagingly to the ordinary psychology of beliefs, desires, and emotions as "folk psychology." The implication is that just as "folk medicine" had to give way to "real medicine," so "folk psychology" will have to give way to a revamped psychology that is grounded in neuroscience. In place of talking cures that address our beliefs, desires, and emotions, tomorrow's healers of the soul will manipulate brain states directly and ignore such outdated categories as beliefs, desires, and emotions.

At least so the story goes. Actual neuroscience research is by contrast a much more modest affair and fails to support materialism's vaulting ambitions. That should hardly surprise us. The neurophysiology of our brains is incredibly plastic and has proven notoriously difficult to correlate with intentional states. For instance, Louis Pasteur, despite suffering a cerebral accident, continued to enjoy a flourishing scientific career. When his brain was examined after he died, it was discovered that half the brain had atrophied. How does one explain a flourishing intellectual life despite a severely damaged brain if mind and brain coincide?

Or consider a more striking example. The December 12, 1980 issue of *Science* contained an article by Roger Lewin titled "Is Your Brain Really Necessary?" In the article, Lewin reported a case study by John Lorber, a British neurologist and professor at Sheffield University:

> "There's a young student at this university," says Lorber, "who has an IQ of 126, has gained a first-class honors degree in mathematics, and is socially completely normal. And yet the boy has virtually no brain." The student's physician at the university noticed that the youth had a slightly larger than normal head, and so referred him to Lorber, simply out of interest. "When we did a brain scan on him," Lorber recalls, "we saw that instead of the normal 4.5-centimeter thickness of brain tissue between the ventricles and the cortical surface, there was just a thin layer of mantle measuring a millimeter or so. His cranium is filled mainly with cerebrospinal fluid."

Against such anomalies, Cabanis's dictum, "the nerves—that's all there is to man," hardly inspires confidence. Yet, as Thomas Kuhn has taught us, a science that is progressing fast and furiously is not about to be derailed by a few anomalies. Neuroscience is a case in point. For all the obstacles it faces in trying to reduce intelligent agency to natural causes, neuroscience persists in the Promethean determination to show that mind does ultimately reduce to neurophysiology. Absent a prior commitment to materialism, this determination will seem misguided. On the other hand, given a prior commitment to materialism, this determination becomes readily understandable.

Understandable yes, obligatory no. Most cognitive scientists do not rest their hopes with neuroscience. Yes, if materialism is correct, then a reduction of intelligent agency to neurophysiology is in principle possible. The sheer difficulty of even attempting this reduction, both experimental and theoretical, however, leaves many cognitive scientists looking for a more manageable field to invest their energies. As it turns out, the field of choice is computer science, and especially its subdiscipline of artificial intelligence. Unlike brains, computers are neat and precise. Also, unlike brains, computers and their programs can be copied and mass-produced. Inasmuch as science thrives on replicability and control, computer science offers tremendous practical advantages over neurological research.

Whereas the goal of neuroscience is to reduce intelligent agency to neurophysiology, the goal of artificial intelligence is to reduce intelligent agency to computation by producing a computational system that equals, or if we are to believe Ray Kurzweil, exceeds human intelligence. Since computers operate deterministically, reducing intelligent agency to computation would indeed constitute a materialistic reduction of intelligent agency. Should artificial intelligence succeed in reducing intelligent agency to computation, cognitive scientists would still have the task of showing in what sense brain function is computational (that is, Marvin Minsky's dictum "the mind is a computer made of meat" would still need to be verified). Even so, the reduction of intelligent agency to computation would go a long way toward establishing a purely materialist basis for human cognition.

An obvious question now arises: Can computation explain intelligent agency? First off, let's be clear that no actual computer system has come anywhere near to simulating the full range of capacities we associate with human intelligent agency. Yes, computers can do certain narrowly circumscribed tasks exceedingly well (like play chess). But require a computer to make a decision based on incomplete information and calling for common sense, and the computer will be lost. Perhaps the toughest problem facing artificial intelligence researchers is what's called the *frame problem*. The frame problem is getting a computer to find the appropriate frame of reference for solving a problem.

Consider, for instance, the following story: A man enters a bar. The bartender asks, "What can I do for you?" The man responds, "I'd like a glass of water." The bartender pulls out a gun and shouts,

"Get out of here!" The man says "thank you" and leaves. End of story. What is the appropriate frame of reference? No, this isn't a story by Franz Kafka. The key item of information needed to make sense of this story is this: The man has the hiccups. By going to the bar to get a drink of water, the man hoped to cure his hiccups. The bartender, however, decided on a more radical cure. By terrifying the man with a gun, the bartender cured the man's hiccups immediately. Cured of his hiccups, the man was grateful and left. Humans are able to understand the appropriate frame of reference for such stories immediately. Computers, on the other hand, haven't a clue.

Ah, but just wait. Give an army of clever programmers enough time, funding, and computational power, and just see if they don't solve the frame problem. Materialists are forever issuing such promissory notes, claiming that a conclusive confirmation of materialism is right around the corner—just give our scientists a bit more time and money. John Polkinghorne refers to this practice as "promissory materialism."

What to do? To refuse such promissory notes provokes the charge of obscurantism, but to accept them means embracing materialism. It is possible to reject promissory materialism without meriting the charge of obscurantism. The point to realize is that a promissory note need only be taken seriously if there is good reason to think that it can be paid. The artificial intelligence community has offered no compelling reason for thinking that it will ever solve the frame problem. Indeed, computers that employ common sense to determine appropriate frames of reference continue utterly to elude computer scientists.

In sum, the empirical evidence for a materialist reduction of intelligent agency is wholly lacking. Indeed, the only thing materialist reductions of intelligent agency have until recently had in their favor is Occam's razor, which has been used to argue that materialist accounts of mind are to be preferred because they are simplest. Yet even Occam's razor, that great materialist mainstay, is proving small comfort these days. Specifically, recent developments in the theory of intelligent design are providing principled grounds against the reduction of intelligent agency to natural causes (cf. my book *The Design Inference*, Cambridge University Press, 1998).

If Not Machines . . .

Until now I've argued that attributing spirituality to machines entails an impoverished view of spirituality, and that the empirical evidence doesn't confirm that machines can bring about minds. But if not machines, what then? What else could mind be except an effect of matter? Or, to restate the question in a more contemporary idiom, what else could mind be except a functional capacity of a complex physical system? It's not that scientists have traced the workings of the brain and discovered how brain states induce mental states. It's rather that scientists have run out of places to look, and that matter seems the only possible redoubt for mind.

The only alternative to a materialist conception of mind appears a Cartesian dualism of spiritual substances that interact preternaturally with material objects. We are left either with a sleek materialism that derives mind from matter or a bloated dualism that makes mind a substance separate from matter. Given this choice, almost no one these days opts for substance dualism. Substance dualism offers two fundamentally different substances, matter and spirit, with no coherent means of interaction. Hence the popularity of reducing mind to matter.

But the choice between materialism and substance dualism is ill-posed. Both materialism and substance dualism are wedded to the same defective view of matter. Both view matter as primary and law-governed. This renders materialism self-consistent since it allows matter to be conceived mechanistically. On the other hand, it renders substance dualism incoherent since undirected natural laws provide no opening for the activity of spiritual substances. But the problem in either case is that matter ends up taking precedence over concrete things. We do not have knowledge of matter but of *things*. As Bishop Berkeley rightly taught us, matter is always an abstraction. Matter is what remains once we remove all the features peculiar to a thing.

Consequently, matter becomes stripped not only of all empirical particularity, but also of any substantial form that would otherwise order it and render it intelligible.

The way out of the materialism-dualism dilemma is to refuse the artificial world of matter governed by natural laws and return to the real world of things governed by the principles appropriate to them. These principles may include natural laws, but they need hardly be coextensive with them. Within this richer world of things as opposed to matter, natural laws lose their status as absolutes and become flexible regularities subject to principles that may be quite distinct from natural laws (principles like intelligent agency).

Within this richer world of things as opposed to matter, the obsession to seek mind in matter quickly dissipates. According to materialism (and here I'm thinking specifically of the scientific materialism that currently dominates Western thought), the world is fundamentally an interacting system of mindless entities (be they particles, strings, fields, or whatever). Accordingly, the only science for studying mind becomes an atomistic, reductionist, and mechanistic science of particles or other mindless entities, which then need to be built up to ever greater orders of complexity by equally mindless principles of association known as natural laws (even the widely-touted "laws of self-organization" fall in here). But the world is a much richer place than materialism allows, and there is no reason to saddle ourselves with its ontology.

The great mistake in trying to understand the mind-body problem is to suppose that it is a scientific problem. It is not. It is a problem of ontology (i.e., that branch of metaphysics concerned with what exists). If all that exists is matter governed by natural laws, then humans are machines. If all that exists is matter governed by natural laws together with spiritual substances that are incapable of coherently interacting with matter, then, once again, humans are machines. But if matter is merely an abstraction gotten by removing all the features peculiar to things, then there is no reason to think that that abstraction, once combined with natural laws or anything else for that matter, will entail the recovery of things. And in that case, there is no reason to think that humans are machines.

According to Owen Barfield, what we call the material or the physical is a "dashboard" that mediates the actual things of the world to us. But the mediation is fundamentally incomplete, for the dashboard can only mirror certain aspects of reality, and that imperfectly. Materialism desiccates the things of this world, and then tries to reconstitute them. Materialism is an exercise in resynthesization. But just as a dried piece of fruit can never be returned to its original freshness, so materialism, once it performs its feat of abstraction, can never return the things as they started out.

This is not for want of cleverness on the part of materialists. It is rather that reality is too rich and the mauling it receives from materialism too severe that even the cleverest materialist cannot recover it. Materialism itself is the problem, not the brand of materialism one happens to endorse (be it scientific, ontological, eliminative, reductive, nonreductive, causal, or conceptual—the literature is full of different spins on materialism that are meant to recover reality for us).

Over a hundred years ago William James saw clearly that science would never resolve the mind-body problem. In his *Principles of Psychology* he argued that neither empirical evidence nor scientific reasoning would settle this question. Instead, he foresaw an interminable debate between competing philosophies, with no side gaining a clear advantage. I close with the following passage from his *Principles of Psychology*, which to me epitomizes the present state of cognitive science:

> We are thrown back therefore upon the crude evidences of introspection on the one hand, with all its liabilities to deception, and, on the other hand, upon *a priori* postulates and probabilities. He who loves to balance nice doubts need be in no hurry to decide the point. Like Mephistopheles to Faust, he can say to himself, "*dazu hast du noch eine lange Frist*" [i.e., "you've got a long wait"], for from generation to generation the reasons adduced on both sides will grow more voluminous, and the discussion more refined.

5

Kurzweil's Turing Fallacy

Thomas Ray

12.2.04

T here are numerous directions from which to criticize Kurzweil's proposal for strong AI. In this essay I will focus on his failure to consider the unique nature of the digital medium when discussing artificial intelligence. But before elaborating on this point, I would like briefly to call attention to some other issues.

Psychic Quantum Mechanics

Kurzweil's interpretation of quantum mechanics leads him to the conclusion that "consciousness, matter, and energy are inextricably

Thomas Ray is a tropical biologist and Professor of Zoology, University of Oklahoma, and developer of the digital evolution software, Tierra.

linked." While this is true in the sense that consciousness arises from the interactions of matter and energy, it is not true in the sense that Kurzweil intends it: that quantum ambiguities are not resolved until they are forced to do so by a conscious observer.

Kurzweil's error is most glaringly apparent in his description of the paper output from a quantum computer: "So the page with the answer is ambiguous, undetermined—until and unless a conscious entity looks at it. Then instantly all the ambiguity is retroactively resolved, and the answer is there on the page. The implication is that the answer is not there until we look at it." He makes the same error in describing the evolution of the universe: "From one perspective of quantum mechanics—we could say that any Universe that fails to evolve conscious life to apprehend its existence never existed in the first place."

Kurzweil does not understand that it is the act of measurement that causes the collapse of the wave function, not conscious observation of the measurement. In practice, the collapse is (probably always) caused by a completely unconscious measuring device. Printing of the result on a paper could be such a measuring device. Subsequent conscious observation of the measurement is irrelevant.

This psychic quantum mechanics did not originate with Kurzweil. It has been around for decades, apparently as a way to deal with Schrödinger's cat. Thus, Kurzweil may be able to point to physicists who hold this view. Similarly, I could point to biologists who believe in the biblical story of creation rather than evolution. The existence of experts who believe a doctrine, however, is no argument for the truth of the doctrine.

Colloquial Chaos

Kurzweil's suggestion that in a process, the time interval between salient events expands or contracts along with the amount of chaos ("the law of time and chaos"), is quite interesting. Yet, the definitions of "salient events" and "chaos" are quite subjective, making the "law" difficult to support. Technically, it would probably be more appropriate to use the word "entropy" in place of "chaos," but for consistency, I will also use "chaos" in this discussion.

Most striking is the apparently inconsistent use of chaos. He states that in an evolutionary process *order* increases, and he says: "Evolution draws upon the chaos in the larger system in which it takes place for its options for diversity." Yet he states that in the development of an individual organism *chaos* increases, and he says: "The development of an organism from conception as a single cell through maturation is a process moving toward greater diversity and thus greater disorder." Kurzweil suggests that in evolution, diversity implies order, while in development, diversity implies disorder.

Through evolution, the diversity of species on Earth has increased, and through development, the diversity of cell types increases. I would characterize both as processes that generate order. Why does Kurzweil think that development generates chaos? His apparent reason is to make his law of time and chaos consistent with our perception of time: Our subjective unit of time grows with our age.

I believe that the scientific community would generally agree that the developmental process up to the period of reproduction is a process of increasing order. In humans, who live well beyond their reproductive years, the condition of the body begins to deteriorate after the reproductive years, and this senescence would generally be considered a process of increasing chaos.

In an effort to fit development seamlessly into his law of time and chaos, Kurzweil presents the whole life cycle from conception to death, as unidirectional, towards increasing chaos. This position is indefensible. The developmental process directly contradicts the law of time and chaos. Development is a process in which the time between salient events increases with order.

He attempts to be clear and concrete in his use of the term chaos: "If we're dealing with the process of evolution of life-forms, then chaos represents the unpredictable events encountered by organisms, and the random mutations that are introduced in the genetic code." He explains: "Evolution draws upon the great chaos in its midst— the ever increasing entropy governed by the flip side of the Law of

Time and Chaos—for its options for innovation." This implies that unpredictable events and mutations are becoming more frequent, a position that would be difficult to defend. His argument is that increasing rates of mutations and unpredictable events are, in part, driving the increasing frequency of "salient events" in evolution. He does not provide any support for this highly questionable argument.

Despite his attempt to be precise, his use of "chaos" is vernacular: "When the entire Universe was just a 'naked' singularity . . . there was no chaos." "As the Universe grew in size, chaos increased exponentially." "Now with billions of galaxies sprawled out over trillions of light-years of space, the Universe contains vast reaches of chaos . . ." "We start out as a single fertilized cell, so there's only rather limited chaos there. Ending up with trillions of cells, chaos greatly expands." It seems that he associates chaos with size, a very unconventional use of the term.

His completely false interpretation of quantum mechanics, his vague and inconsistent use of terms such as "chaos" and "salient events," and his failure to understand the thermodynamics of development represent errors in the basic science from which he constructs his view of the world. These misunderstandings of basic science seriously undermine the credibility of his arguments.

I am not comfortable with the equation of technological development and evolution. I think that most evolutionary biologists would consider these to be quite separate processes, yet, their equation represents a point of view consistent with Kurzweil's arguments and also consistent with the concept of "meme" developed by the evolutionary biologist Richard Dawkins.

Intelligence in the Digital Medium

The primary criticism that I wish to make of Kurzweil's book, however, is that he proposes to create intelligent machines by copying human brains into computers. We might call this the Turing Fallacy. The Turing Test suggests that we can know that machines have become intelligent when we cannot distinguish them from human, in free conversation over a teletype. The Turing Test is one of the biggest red-herrings in science.

It reminds me of early cinema when we set a camera in front of a stage and filmed a play. Because the cinema medium was new, we really didn't understand what it is and what we can do with it. At that point we completely misunderstood the nature of the medium of cinema. We are in almost the same position today with respect to the digital medium.

Over and over again, in a variety of ways, we are shaping cyberspace in the form of the 3D material space that we inhabit. But cyberspace is not a material space and it is not inherently 3D. The idea of downloading the human mind into a computer is yet another example of failing to understand and work with the properties of the medium. Let me give some other examples and then come back to this.

I have heard it said that cyberspace is a place for the mind, yet we feel compelled to take our bodies with us. 3D virtual worlds and avatars are manifestations of this. I have seen virtual worlds where you walk down streets lined by buildings. In one I saw a Tower Records store, whose front looked like the real thing. You approached the door, opened it, entered, and saw rows of CDs on racks and an escalator to take you to the next floor. Just Like The Real Thing!

I saw a demo of Alpha World, built by hundreds of thousands of mostly teenagers. It was the day after Princess Diana died, and there were many memorials to her, bouquets of flowers by fountains, photos of Diana with messages. It looked Just Like The Real memorials to Diana.

I wondered, why do these worlds look and function as much as possible like the real thing? This is cyberspace, where we can do anything. We can move from point A to point B instantly without passing through the space in between. So why are we forcing ourselves to walk down streets and halls and to open doors?

Cyberspace is not a 3D Euclidean space. It is not a material world. We are not constrained by the same laws of physics, unless we impose them upon ourselves. We need to liberate our minds from what we are familiar with before we can use the full potential of cyberspace. Why should we compute collision avoidance for avatars in virtual worlds when we have the alternative to find out how many avatars can dance on the head of a pin?

The WWW is a good counter-example, because it recognizes that in cyberspace it doesn't matter where something is physically located. Amazon.com is a good alternative to the mindlessly familiar 3D Tower Record store.

Let me come back to Kurzweil's ideas on AI. Kurzweil states that it is "ultimately feasible" to:

> . . . scan someone's brain to map the locations, interconnections, and contents of the somas, axons, dendrites, presynaptic vesicles, and other neural components. Its entire organization could then be re-created on a neural computer of sufficient capacity, including the contents of its memory . . . we need only to literally copy it, connection by connection, synapse by synapse, neurotransmitter by neurotransmitter.

This passage most clearly illustrates Kurzweil's version of the Turing Fallacy. It is not only infeasible to "copy" a complex organic organ into silicon without losing its function, but it is the least imaginative approach to creating an AI. How do we copy a seratonin molecule or a presynaptic vesicle into silicon? This passage of the book does not explicitly state whether he is proposing a software simulation from the molecular level up, of a copy of the brain, or if he is proposing the construction of actual silicon neurons, vesicles, neurotransmitters, and their wiring together into an exact copy of a particular brain. Yet in the context of the preceding discussion, it appears that he is proposing the latter.

Such a proposal is doomed to failure. It would be a fantastic task to map the entire physical, chemical, and dynamic structure of a brain. Even if this could be accomplished, there would be no method for building a copy. There is no known technology for building complexly differentiated microscopic structures on such a large scale. If a re-construction method existed, we might expect that a copy made of the same materials, carbon chemistry, if somehow jump-started into the proper dynamic activity, would have the same function (though such a copied brain would require a body to support it). But a copy made of metallic materials could not possibly have the same function. It would be a fantastically complex and intricate dynamic sculpture, whose function would bear no relation to a human brain. And what of the body and its essential sensory integration with the brain?

In order for the metallic "copy" to have the same function, we would have to abstract the functional properties out of the organic neural elements, and find structures and processes in the new metallic medium that provide identical functions. This abstraction and functional-structural translation from the organic into the metallic medium would require a deep understanding of the natural neural processes, combined with the invention of many computing devices and processes which do not yet exist.

However, Kurzweil has stated that one advantage of the brain-copy approach is that "we don't need to understand all of it; we need only to literally copy it." Yet he is ambivalent on this critical point, adding: "To do this right, we do need to understand what the salient information-processing mechanisms are. Much of a neuron's elaborate structure exists to support its own structural integrity and life processes and does not directly contribute to its handling of information."

The structure and function of the brain or its components cannot be separated. The circulatory system provides life support for the brain, but it also delivers hormones that are an integral part of the chemical information processing function of the brain. The membrane of a neuron is a structural feature defining the limits and integrity of a neuron, but it is also the surface along which depolarization propagates signals. The structural and life-support functions cannot be separated from the handling of information.

The brain is a chemical organ, with a broad spectrum of chemical communication mechanisms ranging from microscopic packets of neurotransmitters precisely delivered at target synapses, to nitrogen oxide gas and hormones spread through the circulatory system or diffusing through the intercellular medium of the brain. There also exist a wide range of chemical communications systems with intermediate degrees of specificity of delivery. The brain has evolved its exquisitely subtle and complex functionality based on the properties of these chemical systems. A metallic computation system operates on fundamentally different dynamic properties and could never precisely and exactly "copy" the function of a brain.

The materials of which computers are constructed have fundamentally different physical, chemical, and electrical properties than the materials from which the brain is constructed. It is impossible to create a "copy" of an organic brain out of the materials of computation. This applies not only to the proposition of copying an individual human brain with such accuracy as to replicate a human mind along with its memories, but also to the somewhat less extreme proposition of creating an artificial intelligence by reverse engineering the human brain.

Structures and processes suitable for information processing in the organic medium are fundamentally different from those of the metallic computational medium. Intelligent information processing in the computational medium must be based on fundamentally different structures and processes, and thus cannot be copied from organic brains.

I see three separate processes which are sometimes confounded. Machines having:

1) computing *power* equal to the level of human intelligence

2) computing *performance* equal to the level of human intelligence

3) computing *like* human intelligence

A large portion of Kurzweil's book establishes the first process by extrapolating Moore's Law into the future until individual machines can perform the same number of computations per second as is estimated for the human brain (~2020 A.D.).

I accept that this level of computing power is likely to be reached, someday. But no amount of raw computer power will be intelligent in the relevant sense unless it is properly organized. This is a software problem, not a hardware problem. The organizational complexity of software does not march forward according to Moore's Law.

While I can accept that computing power will inevitably reach human levels, I am not confident that computing performance will certainly follow. The exponential increase of computing power is driven by higher densities and greater numbers of components on chips, not by exponentially more complex chip designs.

The most complex of artifacts designed and built by humans are much less complex that living organisms. Yet the most complex of our creations are showing alarming failure rates. Orbiting satellites and telescopes, space shuttles, interplanetary probes, the Pentium chip, computer operating systems, all seem to be pushing the limits of what we can effectively design and build through conventional approaches.

It is not certain that our most complex artifacts will be able to increase in complexity by an additional one, two or more orders of magnitude, in pace with computing power. Our most complex software (operating systems and telecommunications control systems) already contains tens of millions of lines of code. At present it seems unlikely that we can produce and manage software with hundreds of millions or billions of lines of code. In fact there is no evidence that we will ever be able to design and build intelligent software.

This leads to the next distinction, which is central to my argument, and requires some explanation:

2) computing *performance* equal to the level of human intelligence

3) computing *like* human intelligence

A machine might exhibit an intelligence identical to and indistinguishable from humans, a Turing AI, or a machine might exhibit a fundamentally different kind of intelligence, like some science fiction alien intelligence. I expect that intelligences which emerge from the digital and organic media will be as different as their respective media, even if they have comparable computing performance.

Everything we know about life is based on one example of life, namely, life on earth. Everything we know about intelligence is based on one example of intelligence, namely, human intelligence. This limited experience burdens us with preconceptions and limits our imaginations.

Consider this thought experiment:

We are all robots. Our bodies are made of metal and our brains of silicon chips. We have no experience or knowledge of carbon-based life, not even in our science fiction. Now one of us robots comes to an AI discussion with a flask of methane, ammonia, hydrogen, water, and some dissolved minerals. The robot asks: "Do you suppose we could build a computer from this stuff?"

The engineers among us might propose nano-molecular devices with fullerene switches, or even DNA-like computers. But I am sure they would never think of neurons. Neurons are astronomically large structures compared to the molecules we are starting with.

Faced with the raw medium of carbon chemistry, and no knowledge of organic life, we would never think of brains built of neurons, supported by circulatory and digestive systems, in bodies with limbs for mobility, bodies which can only exist in the context of the ecological community that feeds them.

We are in a similar position today as we face the raw medium of digital computation and communications. The preconceptions and limited imagination deriving from our organic-only experience of life and intelligence make it difficult for us to understand the nature of this new medium, and the forms of life and intelligence that might inhabit it.

How can we go beyond our conceptual limits, find the natural form of intelligent processes in the digital medium, and work with the medium to bring it to its full capacity, rather than just imposing the world we know upon it by forcing it to run a simulation of our physics, chemistry, and biology?

In the carbon medium it was evolution that explored the possibilities inherent in the medium, and created the human mind. Evolution listens to the technology that it is embedded in. It has the advantage of being mindless, and therefore devoid of preconceptions, and not limited by imagination.

I propose the creation of a digital nature. A system of wildlife reserves in cyberspace, in the interstices between human colonizations, feeding off of unused CPU-cycles (and permitted a share of our bandwidth). This would be a place where evolution can spontaneously generate complex information processes, free of the demands of human engineers and market analysts telling it what the target applications are.

Digital naturalists can then explore this cyber-nature in search of applications for the products of digital evolution in the same way that our ancestors found applications among the products of organic nature such as: rice, wheat, corn, chickens, cows, pharmaceuticals, silk, mahogany. But, of course, the applications that we might find in the living digital world would not be material; they would be information processes.

It is possible that out of this digital nature there might emerge a digital intelligence, truly rooted in the nature of the medium, rather than brutishly copied and downloaded from organic nature. It would be a fundamentally alien intelligence, but one which would complement rather than duplicate our talents and abilities.

I think it would be fair to say that the main point of Kurzweil's book is that artificial entities with intelligence equal to and greater than humans will inevitably arise, in the near future. While his detailed explanation of how this might happen focuses on what I consider to be the Turing Fallacy, that is, that it will initially take a human form, Kurzweil would probably be content with any route to these higher intelligences, Turing or non-Turing.

While I feel that AIs must certainly be non-Turing—unlike human intelligences—I feel ambivalent about whether they will emerge at all. It is not the certainty that Kurzweil paints, like the inexorable march of Moore's Law. Raw computing power is not intelligence. Our ability ever to create information processes of a complexity comparable to the human mind is completely unproven and absolutely uncertain.

I have suggested evolution as an alternate approach to producing intelligent information processes. These evolved AIs would certainly be non-Turing AIs. Yet evolution in the digital medium remains a process with a very limited record of accomplishments. We have been able to establish active evolutionary processes, by both natural and artificial selection in the digital medium. But the evolving entities have always contained at most several thousand bits of genetic information.

We do not yet have a measure on the potential of evolution in this medium. If we were to realize a potential within several orders of magnitude of that of organic evolution, it would be a spectacular success. But if the potential of digital evolution falls ten orders of magnitude below organic evolution, then digital evolution will lose its luster. There is as yet no evidence to suggest which outcome is more likely.

The hope for evolution as a route to AI is not only that it would produce an intelligence rooted in and natural to the medium, but that evolution in the digital medium is capable of generating levels of complexity comparable to what it has produced in the organic medium. Evolution is the only process that is proven to be able to generate such levels of complexity. That proof, however, is in the organic rather than the digital medium. Like an artist who can express his creativity in oil paint but not stone sculpture, evolution may be capable of magnificent creations in the organic medium but not the digital.

Yet the vision of the digital evolution of vast complexity is still out there, waiting for realization or disproof. It should encourage us, although we are at the most rudimentary level of our experience with evolution in the digital medium. Nevertheless, the possibilities are great enough to merit a serious and sustained effort.

6

Locked in His Chinese Room: Response to John Searle

Ray Kurzweil

11.10.04

Those Who Build Chinese Rooms are Doomed to Live in Them

John Searle is popular among his followers for what they believe is a staunch defense of the deep mystery of human consciousness against trivialization by strong AI reductionists like Ray Kurzweil. And even though I have always found Searle's logic in his celebrated Chinese Room Argument to be hopelessly tautological, even I had expected him to articulate an elevating treatise on the paradoxes of consciousness. Thus it is with some surprise that I find Searle writing statements such as:

> [H]uman brains cause consciousness by a series of specific neurobiological processes in the brain.
> The essential thing is to recognize that consciousness is a biological process like digestion, lactation, photosynthesis, or mitosis . . .

> The brain is a machine, a biological machine to be sure, but a machine all the same. So the first step is to figure out how the brain does it and then build an artificial machine that has an equally effective mechanism for causing consciousness.

> We know that brains cause consciousness with specific biological mechanisms . . .

So who is being the reductionist here? Searle apparently expects that we can measure the subjectivity of another entity as readily as we measure the oxygen output of photosynthesis.

I will return to this central issue, but I also need to point out the disingenuous nature of many of Searle's quotations and characterizations. For example, he leaves out critical words that dramatically alter the meaning of a statement. For example, Searle writes in his chapter in this book:

> [Ray Kurzweil] insists that they [the machines] will claim to be conscious . . . and consequently their claims will be largely accepted. People will eventually just come to accept without question that machines are conscious. But this misses the point. I can already program my computer so that it says that it is conscious—i.e., it prints out "I am conscious"—and a good programmer can even program it so that it will carry on a rudimentary argument to the effect that it is conscious. But that has nothing to do with whether or not it really is conscious.

Searle fails to point out that I make exactly the same point, and further that I refer not to such idle claims that are easily feasible today but rather to the *convincing* claims of future machines. As one example of many, I write in my book (p. 60) that these claims "won't seem like a programmed response. The machines will be earnest and convincing." but they still are and that's the difference.

Searle writes that I "frequently cite IBM's Deep Blue as evidence of superior intelligence in the computer." The opposite is the case: I cite Deep Blue to (p. 289) "examine the human and [contemporary] machine approaches to chess . . . not to belabor the issue of chess, but rather because [they] *illustrate a clear contrast.*" Human thinking follows a very different paradigm. Solutions emerge in the human brain from the unpredictable interaction of millions of simultaneous self-organizing chaotic processes. There are profound advantages to the human paradigm: we can recognize and respond to extremely subtle patterns. But we can build machines the same way.

Searle states that my book "is an extended reflection of the implications of Moore's Law." But the exponential growth of computing power is only a small part of the story. As I repeatedly state, adequate computational power is *a necessary but not sufficient condition* to achieve human levels of intelligence. Searle essentially doesn't mention my primary thesis: We are learning how to organize these increasingly formidable resources by reverse engineering the human brain itself. By examining brains in microscopic detail, we will be able to recreate and then vastly extend these processes. As I point out below, we have made substantial progress in this endeavor just in the brief period of time since my book was published.

Searle is best known for his "Chinese Room" analogy and has presented various formulations of it over twenty years (see below). His descriptions illustrate a failure to understand the essence of either brain processes or the nonbiological processes that could replicate them. Searle starts with the assumption that the "man" in the room doesn't understand anything because, after all, "he is just a computer," thereby illuminating Searle's own bias. Searle then concludes—no surprise—that the computer doesn't understand. Searle combines this tautology with a basic contradiction: The computer doesn't understand Chinese, yet (according to Searle) can convincingly answer questions in Chinese. But if an entity—biological or otherwise—really doesn't understand human language, it will quickly be unmasked by a competent interlocutor. In addition, for the program to convincingly respond, it would have to be as complex as a human brain. The observers would long be dead while the man in the room spends millions of years following a program billions of pages long.

Most importantly, the man is acting only as the central processing unit, a small part of a system. While the man may not see it, the understanding is distributed across the entire pattern of the program itself and the billions of notes he would have to make to follow the program. *I understand English, but none of my neurons do.* My understanding is represented in vast patterns of neurotransmitter strengths, synaptic clefts, and interneuronal connections. Searle appears not to understand the significance of distributed patterns of information and their emergent properties.

Searle writes that I confuse a simulation for a recreation of the real thing. What my book (and chapter in this book) actually talk about is a third category: functionally equivalent recreation. He writes that we could not stuff a pizza into a computer simulation of the stomach and expect it to be digested. But we could indeed accomplish this with a properly designed artificial stomach. I am not talking about a mere "simulation" of the human brain as Searle construes it, but rather functionally equivalent recreations of its causal powers. As I pointed out, we already have functionally equivalent replacements of portions of the brain to overcome such disabilities as deafness and Parkinson's disease.

Searle writes: "It is out of the question . . . to suppose that . . . the computer is conscious." Given this assumption, Searle's conclusions to the same effect are hardly a surprise. Searle would have us believe that you can't be conscious if you don't possess some specific (albeit unspecified) biological process. No entities based on functionally equivalent processes need apply. This biology-centric view of consciousness is likely to go the way of other human-centric beliefs. In my view, we cannot penetrate the ultimate reality of subjective experience with objective measurement, which is why many classical methods, including Searle's materialist approach, quickly hit a wall.

The Intuitive Linear View Revisited

Searle's slippery and circular arguments aside, nonbiological entities, which today have many narrowly focused skills, are going to vastly expand in the breadth, depth, and subtlety of their intelligence and creativity. Early in his chapter, Searle makes clear his discomfiture with the radical nature of the twenty-first century technologies that I have described and their impact on society. Searle clearly expects the twenty-first century to be much like the twentieth century, and considers any significant deviation from present norms to be absurd on their face. Not once, but twice he expresses incredulity at the notion of virtual sex, for example: "The section on prostitute is a little puzzling to me. . . . But why pay, if it is all an electrically generated fantasy anyway?"

Searle obviously misses the point of virtual reality. Virtual reality is not fantasy; it is a communication medium between two or more people. We already have auditory virtual reality; it's called the telephone. Indeed, that is exactly how the telephone was viewed when it was introduced in the late nineteenth century. People found it remarkable that you could actually "be with" someone else, at least as far as the auditory sense was concerned, despite the fact that you were geographically disparate. And indeed we have a form of sex over phone lines, not very satisfying to many perhaps, but keep in mind it involves only one sense organ. The paradigm, however, is just this: two people communicating, and in some cases one of those persons may be paid for their services. Technology to provide full immersion *visual* shared environments is now being developed, and will be ubiquitous by the end of this decade (with images written directly to our retinas by our eyeglasses and contact lenses). Then, in addition to talking, it will really appear like you are with that other person. As for touching one another, the tactile sense will not be full immersion by the end of this decade, but full immersion virtual shared environments incorporating the auditory, visual, and tactile senses will become available by around 2020. The design of such technology can already be described. When nanobot-based virtual reality becomes feasible around 2030, then these shared environments will encompass all of the senses.

Virtual sex and virtual prostitution are among the more straightforward scenarios for applying full immersion communication technologies, so it is puzzling to me that Searle consistently cites these as among the most puzzling to him. Clearly Searle's thinking about the future is limited by what I referred to in my chapter as the "intuitive linear" view, despite the fact that both he and I have been around long enough to witness the acceleration inherent in the historically accurate exponential view of history and the future.

Twenty-First Century Machine Intelligence Revisited

Beyond Searle's circular, tautological, and often contradictory reasoning, he essentially fails to even address the key points in my chapter and my book, so it is worthwhile reviewing my primary reasoning in my own words. My message concerns the emergence early in the next century of nonbiological entities with enormously powerful intellectual skills and abilities and the profound impact this will have on human society. The primary themes are:

> (1) The power of computer technology per unit cost is growing exponentially. This has been true for the past one hundred years, and will continue well into the next century.

> (2) New hardware technologies such as nanotube-based circuits, which allow three-dimensional computer circuits to be constructed, are already working in laboratories. Such three-dimensional circuits will

ultimately provide physically small devices that vastly exceed the memory and computational ability of the human brain.

(3)   In addition to computation, there is comparable exponential growth in communication, brain scanning, neuron modeling, brain reverse engineering, miniaturization of technology, and many other areas.

(4)   Sufficient computational power by itself is not enough. Adequate computational (and communication) resources are a necessary but not sufficient condition to match the breadth, depth, and subtlety of human capabilities. The organization, content, and embedded knowledge of these resources (i.e., the "software" of intelligence) is also critical.

(5)   A key resource for understanding and ultimately recreating the software of intelligence is the human brain itself. By probing the human brain, we are already learning its methods. We are already applying these types of insights (e.g., the front-end sound-wave transformations used in automatic speech recognition systems are based on early auditory processing in mammalian brains). The brain is not invisible to us. Our ability to scan and understand human neural functioning both invasively and noninvasively is scaling up exponentially.

(6)   We have already created detailed replications of substantial neuron clusters. These replications (not to be confused with the simplified mathematical models used in many contemporary "neural nets") recreate the highly parallel analog-digital functions of these neuron clusters, and such efforts are also scaling up

exponentially. This has nothing to do with manipulating symbols, but is a detailed and realistic recreation of what Searle refers to as the "causal powers" of neuron clusters. Human neurons and neuron clusters are certainly complicated, but their complexity is not beyond our ability to understand and recreate using other mediums. I cite specific recent progress below.

(7)   We've already shown that the causal powers of substantial neuron clusters cannot only be recreated, but actually placed in the human brain to replace disabled brain portions. These are not mere simulations, but functionally equivalent recreations of the causal powers of neuron clusters.

(8)   With continuing exponential advances in computer hardware, neuron modeling, and human brain scanning and understanding, it is a conservative statement to say that we will have detailed models of neurons and complete maps of the human brain within thirty years that enable us to reverse engineer its organization and content. This is no more startling a proposition than was the proposal to scan the entire human genome 14 years ago. Well before that, we will have nonbiological hardware with the requisite capacity to replicate its causal powers. Human brain level computational power, together with an understanding of the organization and content of human intelligence gained through such reverse engineering efforts, will be a formidable combination.

(9)   Although contemporary computers can compete with human intelligence in narrow domains (e.g., chess, diagnosing blood cell images, recognizing land terrain images in a cruise missile, making financial

investment decisions), their overall intelligence lacks the subtlety and range of human intelligence. Compared to humans, today's machines appear brittle and formulaic. But contemporary computers are still a million times simpler than the human brain. The depth and breadth of the behavior of nonbiological entities will appear quite different when the million-fold difference in complexity is reversed, and when we can apply powerful models of biological processes.

(10) There are profound advantages to nonbiological intelligence. If I spend years learning French, I can't transfer that knowledge to you. You have to go through a similar painstaking process. We cannot easily transfer (from one person to another) the vast pattern of neurotransmitter strengths, synaptic clefts, and other neural elements that represents our human knowledge. But we won't leave out quick downloading ports in our nonbiological recreations of neuron clusters. Machines will be able, therefore, to rapidly share their knowledge.

(11) Virtual personalities can claim to be conscious today, but such claims are not convincing. They lack the subtle and profound behavior that would make such claims compelling. But the claims of nonbiological entities some decades from now—entities that are based on the detailed design of human thinking—will not be so easily dismissed.

(12) The emergence of highly advanced intelligence in our machines will have a profound impact on all aspects of our human-machine civilization.

Recent Progress in Brain Reverse Engineering

Critical to my thesis is the issue of brain reverse engineering, so it is worth commenting on recent progress in this area. Just in the two years since my recent book was published, progress in this area has been remarkably fast. The pace of brain reverse engineering is only slightly behind the availability of the brain scanning and neuron structure information. There are many contemporary examples, but I will cite just one, which is a comprehensive model of a significant portion of the human auditory processing system that Lloyd Watts <www.lloydwatts.com> has developed from both neurobiology studies of specific neuron types and brain interneuronal connection information. Watts' model includes more than a dozen specific brain modules, five parallel paths and includes the actual intermediate representations of auditory information at each stage of neural processing. Watts has implemented his model as real-time software which can locate and identify sounds with many of the same properties as human hearing. Although a work in progress, the model illustrates the feasibility of converting neurobiological models and brain connection data into working functionally equivalent recreations. Also, as Hans Moravec and others have speculated, these efficient machine implementations require about 1,000 times less computation than the theoretical potential of the biological neurons being recreated.

The brain is not one huge "tabula rasa" (i.e., undifferentiated blank slate), but rather an intricate and intertwined collection of hundreds of specialized regions. The process of "peeling the onion" to understand these interleaved regions is well underway. As the requisite neuron models and brain interconnection data becomes available, detailed and implementable models such as the auditory example above will be developed for all brain regions.

On the Contrast Between
Deep Blue and Human Thinking

To return to Searle's conceptions and misconceptions, he misconstrues my presentation of Deep Blue. As I mentioned above, I discuss Deep Blue because it illustrates a clear contrast between this particular approach to building machines that perform certain structured tasks such as playing chess, and the way that the human brain works. In my book, I use this discussion to present a proposal to build these systems in a different way—a more human way (see below). Searle concentrates entirely on the methods used by Deep Blue, which completely misses the point.

Searle's chapter is replete with misquotations. For example, Searle states:

> So what, according to Kurzweil and Moore's Law, does the future hold for us? We will very soon have computers that vastly exceed us in intelligence. Why does increase in computing power automatically generate increased intelligence? Because intelligence, according to Kurzweil, is a matter of getting the right formulas in the right combination and then applying them over and over, in his sense "recursively," until the problem is solved.

This is a completely erroneous reference. I repeatedly state that increases in computing power do not automatically generate increased intelligence. Furthermore, with regard to Searle's reference to recursion, I present the recursive method as only one technique among many, and as a method suitable only for a narrow class of problems such as playing board games. I never present this simple approach as the way to create human-level intelligence in a machine.

If you read Searle's chapter and do not read my book, you would get the impression that I present the method used by Deep Blue as the ultimate paradigm for machine intelligence. It makes me wonder if Searle actually read the book, or just selectively picked phrases out of context. I repeatedly contrast the recursive methods of Deep Blue with the pattern recognition based paradigm used by the human brain. The field of pattern recognition represents my own technical area of expertise. Human pattern recognition is based on a paradigm in which solutions emerge from the interplay of many interacting processes (see below). What I clearly describe in the book is moving away from the formulaic approaches used by many contemporary AI systems and moving towards the human paradigm of pattern recognition.

Searle's explanation of how Deep Blue works is essentially correct (thanks in large measure to my explaining it to him in response to his urgent email messages to me asking me to clarify for him how Deep Blue works). Although the basic recursive method of rapidly expanding move-countermove sequences is simple, the evaluation at the "leaves" of this move-countermove tree (the scoring function) is really the heart of the method. If you have a simple scoring function, then the method is indeed simple and dependent merely on brute force in computational speed. However, the scoring function is not necessarily simple. Deep Blue's scoring function uses up to 8,000 different features, and is more complex than most.

Deep Blue is able to consider billions of board situations and creates an enormous tree of move-countermove possibilities. Since our human neurons are so slow (at least ten million times slower than electronic circuits), we only have time to consider at most a few hundred board positions. Since we are unable to consider the billions of move-countermove situations that a computer such as Deep Blue evaluates, what we do instead is to "deeply" consider each of these situations. So how do we do that? By using *pattern recognition*, which is the heart of human intelligence. We have the ability to recognize situations as being similar to ones we have thought about previously. A chess master such as Kasparov will have mastered up to one hundred thousand such board situations. As he plays, he recognizes situations as being similar to ones he has thought about before and then calls upon his memory of those previous thoughts (e.g., "this is just like that situation that I got into three years ago against grandmaster so-and-so when I forgot to protect my trailing pawn . . .").

I discuss this in my book in order to introduce a proposal to build game-playing machines in a new and hybrid way which would combine the current strength of machines (i.e., the ability to quickly sift through a vast combinatorial explosion of move-countermove sequences) with the more human-like pattern recognition paradigm which represents at least a current superiority of human thinking. Basically, the idea is to use a large (machine-based) neural net to replace the scoring function. Prior to playing, we train that neural net on millions of examples of real-world chess playing (or whatever other game or problem we are addressing). With regard to chess, we have most of the master games of this century on-line, so we can train this extensive neural net on every master game. And then instead of just using an arbitrary set of rules or procedures at the terminal leaves (i.e., the scoring function), we would use this fully trained neural net to make these evaluations. This would combine the combinatorial approach with a pattern recognition approach (which, as I mentioned above, is my area of technical expertise).

I proposed this to Murray Campbell, head of the IBM Deep Blue team, and he was very interested in the idea, and we were going to pursue it, but then IBM cancelled the Deep Blue project. I may yet return to the idea. Recently I brought up the idea again with Campbell.

Searle completely misconstrues this discussion in my book. It is not at all my view that the simple recursive paradigm of Deep Blue is exemplary of how to build flexible intelligence in a machine. The pattern recognition paradigm of the human brain is that solutions emerge from the chaotic and unpredictable interplay of millions of simultaneous processes. And these pattern recognizers are themselves organized in elaborate and shifting hierarchies. In contrast to today's computers, the human brain is massively parallel, combines digital and analog methods, and represents knowledge as highly distributed patterns encoded in trillions of neurotransmitter strengths.

A failure to understand that computing processes are capable of being—just like the human brain—chaotic, unpredictable, messy, tentative, and emergent is behind much of the criticism of the prospect of intelligent machines that we hear from Searle and other essentially materialist philosophers. Inevitably, Searle comes back to a criticism of "symbolic" computing: that orderly sequential symbolic processes cannot recreate true thinking. I think that's true.

But that's not the only way to build machines, or computers.

So-called computers (and part of the problem is the word "computer" because machines can do more than "compute") are not limited to symbolic processing. Nonbiological entities can also use the emergent self-organizing paradigm, and indeed that will be one great trend over the next couple of decades, a trend well under way. Computers do not have to use only 0 and 1. They don't have to be all digital. The human brain combines analog and digital techniques. For example, California Institute of Technology Professor Carver Mead and others have shown that machines can be built by combining digital and analog methods. Machines can be massively parallel. And machines can use chaotic emergent techniques just as the brain does.

My own background is in pattern recognition, and the primary computing techniques that I have used are not symbol manipulation, but rather self-organizing methods such as neural nets, Markov models, and evolutionary (sometimes called genetic) algorithms.

A machine that could really do what Searle describes in the Chinese Room would not be merely "manipulating symbols" because that approach doesn't work. This is at the heart of the philosophical slight of hand underlying the Chinese Room (but more about the Chinese Room below).

It is not the case that the nature of computing is limited to manipulating symbols. Something is going on in the human brain, and there is nothing that prevents these biological processes from being reverse engineered and replicated in nonbiological entities.

Searle writes that "Kurzweil assures us that Deep Blue was actually thinking." This is one of Searle's many out-of-context quotations. The full quotation from my book addresses diverse ways of viewing the concept of thinking, and introduces my proposal for building Deep Blue in a different, more human way:

After Kasparov's 1997 defeat, we read a lot about how Deep Blue was just doing massive number crunching, not really "thinking" the way his human rival was doing. One could say that the opposite is the case, that Deep Blue was indeed thinking through the implications of each move and countermove; and that it was Kasparov who did not have time to really think very much during the tournament. Mostly he was just drawing upon his mental database of situations he had thought about long ago. Of course, this depends on one's notion of thinking, as I discussed in chapter three. But if the human approach to chess—*neural network based pattern recognition used to identify situations from a library of previously analyzed situations*—is to be regarded as true thinking, then why not program our machines to work the same way? The third way: And that's my idea that I alluded to above as the third school of thought in evaluating the terminal leaves in a recursive search. . . .

Finally, a comment on Searle's view that the "real competition was not between Kasparov and the machine, but between Kasparov and a team of engineers and programmers." Both Deep Blue and Kasparov obtain input and modification to their knowledge bases and strategies from time to time between games. But both Deep Blue and Kasparov use their internal knowledge bases, strategies, and abilities to play each game without any outside assistance or intervention during the game.

On Searle and his Chinese Rooms

John Searle is probably best known for his Chinese Room Argument, which adherents believe demonstrates that machines (i.e., nonbiological entities) can never truly understand anything of significance (such as Chinese). There are several versions of the Chinese Room, of which I will discuss three.

Chinese Room One: A Person and a Computer in a Room

The first involves a person and a computer in a room. I quote here from Professor Searle's 1992 book:

> I believe the best-known argument against strong AI was my Chinese room argument (Searle 1980a) that showed that a system could instantiate a program so as to give a perfect simulation of some human cognitive capacity, such as the capacity to understand Chinese, even though that system had no understanding of Chinese whatever. Simply imagine that someone who understands no Chinese is locked in a room with a lot of Chinese symbols and a computer program for answering questions in Chinese. The input to the system consists in Chinese symbols in the form of questions; the output of the system consists in Chinese symbols in answer to the questions. We might suppose that the program is so good that the answers to the questions are indistinguishable from those of a native Chinese speaker. But all the same, neither the person inside nor any other part of the system literally understands Chinese; and because the programmed computer has nothing that this system does not have, the programmed computer, qua computer, does not understand Chinese either. Because the program is purely formal or syntactical and because minds have mental or semantic contents, any attempt to produce a mind purely with computer programs leaves out the essential features of the mind.

First of all, it is important to recognize that for this system—the person and the computer—to, as Professor Searle puts it, "give a perfect simulation of some human cognitive capacity, such as the capacity to understand Chinese" and to convincingly answer questions in Chinese, this system is essentially passing a Chinese Turing Test. It is entirely equivalent to a Turing Test. In the Turing Test, a computer answers questions in a natural language such as English, or it could be Chinese, in a way that is convincing to a human judge. That is essentially the premise here in the Chinese Room. Keep in mind that we are not talking about answering questions from a fixed list of stock questions (because that's a trivial task), but answering any unanticipated question or sequence of questions from a knowledgeable human interrogator, just as in Turing's eponymous test.

Now, the human in the Chinese Room has little or no significance. He is just feeding things into the computer and mechanically transmitting the output of the computer. And the computer and the human don't need to be in a room either. *Both the human and the room are irrelevant*. The only thing that is significant is the computer.

Now for the computer to really perform this "perfect simulation of a human cognitive capacity, such as the capacity to understand Chinese," it would have to, indeed, understand Chinese. It has, according to the very premise "the capacity to understand Chinese," so it is then entirely contradictory to say that "the programmed computer . . . does not understand Chinese." The premise here directly contradicts itself.

A computer and computer program *as we know them today* could not successfully perform the described task. So if we are to understand the computer to be like today's computers, then it is not fulfilling the premise. The only way that it could fulfill the premise would be for the computer to have the depth and complexity that a human has. That was Turing's brilliant insight in proposing the Turing Test, that convincingly answering questions in a human language really probes all of human intelligence. We're not talking here about answering a question from a canned set of questions, but answering any possible sequence of questions from an intelligent human questioner. A system that could only answer a fixed set of questions would quickly be unmasked by a knowledgeable interlocutor. That requires a human level of intelligence.

A computer that is capable of accomplishing this—a computer that we will run into a few decades from now—will need to be of human complexity or greater, and will indeed understand Chinese in a deep way—because otherwise it would never be convincing in its claim to understand Chinese.

So just stating the computer "does not literally understand Chinese" does not make sense. It contradicts the entire premise.

To claim that the computer is not conscious is not compelling either. To be consistent with some of Searle's other statements, we have to conclude that we really don't know if it is conscious or not. With regard to relatively simple machines, including today's computers, while we can't state for certain that these entities are not conscious, their behavior, including their inner workings, don't give us that impression. But that will not be true for a computer that can really do what is needed in the Chinese room. Such a computer will at least *seem* conscious. Whether it is or not, we really cannot make a definitive statement. But just declaring that it is obvious that the computer (or the entire system of the computer, person and room) is not conscious is far from a compelling argument.

In the quote I read above, Professor Searle is saying that "the program is purely formal or syntactical." But as I pointed out above, that is a bad assumption based on Searle's failure to understand the requirements of such a technology. This assumption is behind much of the criticism of AI that we have heard from certain AI critics such as Searle. A program that is purely formal or syntactical will not be able to understand Chinese, and it won't "give a perfect simulation of some human cognitive capacity."

But again, we don't have to build our machines that way. We can build them the same way nature built the human brain: using chaotic emergent methods that are massively parallel. Furthermore, there is nothing preventing machines from mastering semantics. There is nothing inherent in the concept of a machine that restricts its expertise to the level of syntax alone. Indeed if the machine inherent in Searle's conception of the Chinese Room had not mastered semantics, it would not be able to convincingly answer questions in Chinese and thus would contradict Searle's own premise.

One approach, as I discuss at length in my book and in my chapter in this book, is to reverse engineer and copy the methods of the human brain (with possible extensions). And if it is a Chinese human brain, the copy will understand Chinese. I am not talking about a simulation per se, but rather a duplication of the causal powers of the massive neuron cluster that constitutes the brain, at least those causal powers salient and relevant to thinking.

Will such a copy be conscious? I don't think the Chinese Room Argument tells us anything about this question.

Searle's Chinese Room Argument Can Be Applied to the Human Brain Itself

Although it is clearly not his intent, Searle's own argument implies that the human brain has no understanding. He writes:

> "The computer . . . succeeds by manipulating formal symbols. The symbols themselves are quite meaningless: they have only the meaning we have attached to them. The computer knows nothing of this, it just shuffles the symbols."

Searle acknowledges that biological neurons are machines, so if we simply substitute the phrase "human brain" for "computer" and "neurotransmitter concentrations and related mechanisms" for "formal symbols," we get:

> *The [human brain] . . . succeeds by manipulating [neurotransmitter concentrations and related mechanisms]. The [neurotransmitter concentrations and related mechanisms] themselves are quite meaningless: they have only the meaning we have attached to them. The [human brain] knows nothing of this, it just shuffles the [neurotransmitter concentrations and related mechanisms].*

Of course, neurotransmitter concentrations and other neural details (e.g., interneuronal connection patterns) have no meaning in and of themselves. The meaning and understanding that emerges in the human brain is exactly that: an *emergent* property of its complex patterns of activity. The same is true for machines. Although the "shuffling symbols" do not have meaning in and of themselves, the emergent patterns have the same potential role in nonbiological systems as they do in biological systems such as the brain. As Hans Moravec has written, "Searle is looking for understanding in the wrong places . . . [he] seemingly cannot accept that real meaning can exist in mere patterns." *I would like more on this*

Chinese Room Two: People Manipulating Slips of Paper

Okay, now let's address a second conception of the Chinese Room. In this conception of the Chinese Room Argument, the room does not include a computer but has a room full of people manipulating slips of paper with Chinese symbols on it. The idea is that this system of a room, people, and slips of paper would convincingly answer questions in Chinese, but none of the participants would know Chinese, nor could we say that the whole system really knows Chinese. Not in a conscious way, anyway. Searle then essentially ridicules the idea that this "system" could be conscious. What are we to consider conscious, Searle asks: the slips of paper, the room? Of course the very notion sounds absurd, so the point is made.

One of the problems with this version of the Chinese Room Argument is that this model of the Chinese Room does not come remotely close to really solving the specific problem of answering questions in Chinese. This form of Chinese Room is really a description of a machine-like process that uses the equivalent of a table look-up, with perhaps some straightforward logical manipulations, to answer questions. It would be able to answer some limited number of canned questions. But if it were to answer *any* arbitrary question that it might be asked, this process would really have to understand Chinese in the same way that a Chinese person does. Again, it is essentially being asked to pass a Chinese Turing Test. And as such, it would need to be as clever, and about as complex, as a human brain, a Chinese human brain. And straightforward table look-up algorithms are simply not going to work.

If we want to recreate a brain that understands Chinese using people as little cogs in the recreation, we would really need a person for each neural connection, so we would need about a hundred trillion people, which means about ten thousand planet Earths with ten billion persons each. This would require a rather large room. And even if extremely efficient organized, this system would run many thousands of times slower than the Chinese brain it is attempting to recreate (by the way, I say thousands, and not millions or trillions because the human brain is very slow compared to electronic circuits—200 calculations per second versus about one billion for machines today).

So Professor Searle is taking an utterly unrealistic solution, one that does not come close to fulfilling its own premise, and then asks us to perform a mental thought experiment that considers whether or not this unrealistic system is conscious, or knows anything about Chinese. The very word "room" is misleading, as it implies a limited number of people with some manageable number of slips of papers. So people think of this so-called "room" and these slips of papers and the rules of manipulating the slips of paper and then are asked to consider if this "system" is conscious. The apparent absurdity of considering this simple system to be conscious is therefore supposed to show that such a recreation of an intelligent process would not really "know" Chinese.

However, if we were to do it right, so that it would actually work, it would take on the order of a hundred trillion people. Now it's true that none of these hundred trillion people would need to know anything about Chinese, and none of them would necessarily know what is going on in this elaborate system. But that's equally true of the neural connections in a real human brain. None of the hundred trillion connections in my brain knows anything about this Discovery Institute book chapter I am writing, nor do any of them know English, nor any of the other things that I know. None of them are conscious of this chapter, nor of any of the things I am conscious of. Probably none of them are conscious at all. But the entire system of them, that is Ray Kurzweil, is conscious. At least, I'm claiming that I'm conscious.

So if we scale up Searle's Chinese Room to be the rather massive "room" it needs to be, who's to say that the entire system of a hundred trillion people simulating a Chinese Brain that knows Chinese isn't conscious? Certainly, it would be correct to say that such a system knows Chinese. And we can't say that it is not conscious anymore than we can say that about any other process. We can't know the subjective experience of another entity (and in at least some of Searle's writings, he appears to acknowledge this limitation). And this massive hundred trillion person "room" is an entity. And perhaps it is conscious. Searle is just declaring ipso facto that it isn't conscious, and that this conclusion is obvious. It may seem that way when you call it a room, and talk about a limited number of people manipulating a limited number of pieces of paper. But as I said, such a system doesn't remotely work.

A key to the philosophical sleight of hand implicit in the Chinese Room Argument has specifically to do with the complexity and scale of the system. Searle says that whereas he cannot prove that his typewriter or tape recorder are not conscious, he feels it is obvious that they are not. Why is this so obvious? At least one reason is because a typewriter and a tape recorder are relatively simple entities.

But the existence or absence of consciousness is not so obvious in a system that is as complex as the human brain, indeed one that may be a direct copy of the organization and causal powers of a real human brain. If such a "system" acts human and knows Chinese in a human way, is it conscious? Now the answer is no longer so obvious. What Searle is saying in the Chinese Room Argument is that we take a simple "machine"—and the conception of a room of people manipulating slips of paper is indeed a simple machine—and then consider how absurd it is to consider such a simple machine to be conscious.

I would agree that a simple machine appears not to be conscious, and that a room of people manipulating slips of paper does not appear to be conscious. But such a simple machine, whether it be a typewriter, a tape recorder, or a room of people manipulating slips of paper cannot possibly answer questions in Chinese. So the fallacy has everything to do with the scale and complexity of the system. Simple machines do not appear to be conscious (again, this is not a proof, but a reasonable conclusion nonetheless). The possible consciousness of machines that are as complex as the human brain is an entirely different question. Complexity alone does not necessarily give us consciousness, but the Chinese Room tells us nothing about whether or not such a system is conscious. The way Searle describes this Chinese Room makes it sound like a simple system, so it seems reasonable to conclude that it isn't conscious. What he doesn't tell you is that the room needs to be much bigger than the solar system, so this apparently simple system isn't really so simple at all.

Chinese Room Three: A Person with a Rule Book

A third variant of the Chinese Room is that there is only one person manipulating slips of papers according to a "rule book." Searle then asks what we are we to consider conscious: the slips of paper, the rule book, the room? Again, the humorous absurdity of the situation clearly implies that the system is not conscious, and does not really "know" Chinese.

But again, it would be utterly infeasible for this little system to provide "a perfect simulation of some human cognitive capacity, such as the capacity to understand Chinese" unless the rule book were to be as complex as a human brain that understands Chinese. And then it would take absurdly long for the human to follow the trillions of rules.

Okay, how about if the rule book simply listed every possible question, and then provided the answer? This would be even less feasible, as the number of possible questions is in the trillions of trillions.

Also keep in mind that the answer to a question would need to consider all of the dialogue that came before it.

The term "rule book" implies a book of hundreds or maybe thousands of pages of rules, but not many trillions of pages.

So again we have a simple machine—a person and a "rule book"—and the apparent absurdity of such a simple system "knowing" Chinese or being conscious. But what really is absurd is the notion that such a system, even in theory, could really answer questions in Chinese in a convincing way.

The version of the Chinese Room Searle cites in his chapter in this book is closest to this third conception. One just replaces "rule book" with "computer program." But as I point out above, the man in the room is acting like the central processing unit (CPU) of the computer carrying out the program. One could indeed say that the CPU of a computer, being only a small part of a larger system, does not understand what the entire system understands. One has to look for understanding from the right perspective. The understanding, in this case, is distributed across the entire system, including a vastly complex program, and the billions of little notes that the man would have to keep and organize in order to actually follow the program. That's where the understanding lies, not in the CPU (i.e., the man in the room) alone. It is a distributed understanding embedded in a vast pattern, a type of understanding that Searle appears not to recognize.

Ray Kurzweil's Chinese Room: With Decorations from the Ming Dynasty

Okay, so here is my conception of the Chinese Room. Call it Ray Kurzweil's Chinese Room:

There is a human in a room. The room has decorations from the Ming Dynasty. There is a pedestal on which sits a mechanical typewriter. The typewriter has been modified so that there are Chinese symbols on the keys instead of English letters. And the mechanical linkages have been cleverly altered so that when the human types in a question in Chinese, the typewriter does not type the question, but instead types the answer to the question.

Now the person receives questions in Chinese characters, and dutifully presses the appropriate keys on the typewriter. The typewriter types out not the question, but the appropriate answer. The human then passes the answer outside the room.

So here we have a room with a man in it that appears to know Chinese, yet clearly the human does not know Chinese. And clearly the typewriter does not know Chinese either. It is just an ordinary typewriter with its mechanical linkages modified. So despite the fact that the man in the room can answer questions in Chinese, who or what can we say truly knows Chinese? The decorations?

Now you might have some objections to my Chinese Room.

You might point out that the decorations don't seem to have any significance.

Yes, that's true. Neither does the pedestal. The same can be said for the human, and for the room.

You might also point out that the premise is absurd. *Just changing the mechanical linkages in a mechanical typewriter could not possibly enable it to convincingly answer questions in Chinese* (not to mention the fact that we can't fit all the Kanji symbols on the keys).

Yes, that's a valid objection as well. Now the only difference between my Chinese Room conception, and the several proposed by Professor Searle, is that it is patently obvious in my conception that it couldn't possibly work. It is obvious that my conception is absurd. That is not quite as apparent to many readers or listeners with regard to the Searle Chinese Rooms. However, it is equally the case.

Now, wait a second. We can make my conception work, just as we can make Searle's conceptions work. All you have to do is to make the typewriter linkages as complex as a human brain. And that's theoretically (if not practically) possible. But the phrase "typewriter linkages" does not suggest such vast complexity. The same is true when Searle talks about a person manipulating slips of paper or following a book of rules or a computer program. These are all equally misleading conceptions.

The Chinese Room and Chess

Searle's application of his Chinese Room to chess is equally misleading. He says the man in the room "looks up in a book what he is supposed to do." So again, we have a simple look-up procedure. What sort of book is Searle imagining? If it lists all the chess situations that the man might confront, there wouldn't be enough particles in the Universe to list them all, given the number of possible permutations of chess boards. If, on the other hand, the book contains the program that Deep Blue follows, the man would take thousands of years to make a move, which last time I checked, is not regulation chess.

Searle's primary point is contained in his statement:

> The man understands nothing of chess; he is just a computer. And the point of the parable is this: if the man does not understand chess on the basis of running the chess-playing program, neither does any other computer solely on that basis.

As I pointed out earlier, Searle is simply assuming his conclusion: the man "is just a computer," so obviously (to Searle) he cannot understand anything. But the entire system which includes the rule book and the man following the rule book does "understand" chess, or else it wouldn't be able to play the game.

It should also be pointed out that playing good chess, even championship chess, is a lot easier than convincingly answering questions in a natural human language such as Chinese. But then, Searle shifts the task from playing chess to being knowledgeable about chess in a human context: knowing something about the history and role of chess, having knowledge about the roles of kings and queens who do not necessarily stand on chess squares, having reasons to want to win the game, being able to articulate such reasons, and so on. A reasonable test of such knowledge and understanding of context would be answering questions about chess and engaging in a convincing dialogue (in the Turing Test sense) about chess using a human language such as English or Chinese. And now we have a task that is very similar to the original Chinese Room task, to which my comments above pertain.

On the Difference Between Simulation and Re-Creation

This discussion of Searle's, which he numbers (1), is so hopelessly confused that it is difficult to know where to begin to unravel his tautological and contradictory reasoning.

Let me start with Searle's stomach analogy. He writes:

> What the computer does is a simulation of these processes, a symbolic model of the processes. But the computer simulation of brain processes that produce consciousness stands to real consciousness as the computer simulation of the stomach processes that produce digestion stands to real digestion. You do not cause digestion by doing a computer simulation of digestion. Nobody thinks that if we had the perfect computer simulation running on the computer, we could stuff a pizza into the computer and it would thereby digest it. It is the same mistake to suppose that when a computer simulates the processes of a conscious brain it is thereby conscious.

As I point out in at the beginning of my discussion of Searle's chapter above, Searle confuses simulation with functionally equivalent recreation. We could indeed stuff a pizza into an artificial stomach. It may have a very different design than an ordinary human stomach, but if properly designed, it would digest the pizza as well, or perhaps even better than, a real stomach (in the case of some people's stomachs, that probably wouldn't be so hard to do).

In my chapter and in my book, I discuss the creation of functionally equivalent recreations of individual neurons (which has been done), of substantial clusters of neurons (which has also been done), and, ultimately, of the human brain. I am not talking about conventional neural nets, which involve mathematically simplified neurons, but recreations of the full complexity of the digital-analog behavior and response of human and other mammalian neurons and neuron clusters. And these clusters have been growing rapidly (in accordance with the law of accelerating returns). A few years ago, we could only replicate individual neurons, then we could replicate clusters of tens of neurons, then hundreds, and scientists are now replicating clusters of thousands of neurons. Scaling up to the billions of neurons in the human brain may seem daunting, but so did the human genome scan when first proposed.

I don't assume that a perfect or near-perfect recreation of a human brain would necessarily be conscious. But we can expect that it would exhibit the same subtle, complex behavior and abilities that we associate with humans. Our wonderful ability to connect chessboard kings to historical kings and to reflect on the meaning of chess, and all of our other endearing abilities to put ideas in a panoply of contexts is the result of the complex swirl of millions of interacting processes that take place in the human system. If we recreate (and ultimately, vastly extend) these processes, we will get comparably rich and subtle behavior. Such entities will at least convincingly seem conscious. But I am the first to agree that this does not prove that they are in fact conscious.

Searle writes:

> The computer, as we saw in our discussion of the chess-playing program, succeeds by manipulating formal symbols. The symbols themselves are quite meaningless: they have only the meaning we have attached to them. The computer knows nothing of this; it just shuffles the symbols. And those symbols are not by themselves sufficient to guarantee equivalent causal powers to actual biological machinery like human stomachs and human brains.

Here again, Searle assumes that the methods used by Deep Blue are the only way to build intelligent machines. Searle may assume this, but that is clearly not what my book discusses. There are other methods that do not involve the manipulation of formal symbols in this sense. We have discovered that the behavior and functioning of neurons, while quite complex, are describable in mathematical terms. This should not be surprising, as neurons are constructed of real materials following natural laws. And chips have been created that implement these descriptions, and the chips operate in a very similar manner to biological neurons. We are even putting such chips in human brains to replace disabled portions of those brains, as in the case of neural implants for deafness, Parkinson's Disease, and a growing list of other conditions.

There is nothing in Searle's arguments that argues against our ability to scale up these efforts to capture all of human intelligence, and then extend it in nonbiological mediums. As I pointed out above, these efforts are already scaling up very quickly.

Searle writes:

> Kurzweil points out that not all computers manipulate symbols. Some recent machines simulate the brain by using networks of parallel processors called "neural nets," which try to imitate certain features of the brain. But that is no help. We know from the Church-Turing Thesis, a mathematical result, that any computation that can be carried out on a neural net can be carried out on a symbol-manipulating machine. The neural net gives no increase in computational power. And simulation is still not duplication.

It is remarkable that Searle describes the Church-Turing Thesis as a "mathematical result," but more about that later. Searle here is confusing different results of Church and Turing. Turing and Church independently derived mathematical theorems that show that methods such as a neural net can be carried out, albeit very slowly, on a

Turing Machine, which can be considered as a universal symbol-manipulating machine. They also put forth a conjecture, which has become known as the Church-Turing Thesis, which is not mathematical in nature, but rather relates certain abilities of the human brain (in particular its mathematical abilities) to abilities of a Turing Machine.

We know in practical terms that we can precisely replicate neural functioning in electronic devices. No one has demonstrated any practical limits to our ability to do this. In the book, I discuss our efforts to understand the human brain, and the many different schools of thought pursuing the replication of its abilities.

Searle acknowledges that neural nets can be emulated through computation. Well, that only confirms my thesis. Although many contemporary neural nets involve highly simplified models of neurons, a neural net does not necessarily need to be based on such simplified models of biological neurons. They can be built from models of neurons that are just as complex as biological neurons, or even more complex. And doing so would not change the implications of Turing's and Church's theorems. So we could still replicate these neural nets through forms of computation. And indeed we have been successfully doing exactly this, and such efforts are rapidly increasing in complexity.

As for simulation not being duplication, as I pointed out above, I am specifically talking about functionally equivalent duplication.

Searle writes:

> He [Kurzweil] does not claim to know that machines will be conscious, but he insists that they will claim to be conscious, and will continue to engage in discussions about whether they are conscious, and consequently their claims will be largely accepted. People will eventually just come to accept without question that machines are conscious.

> But this misses the point. I can already program my computer so that it says that it is conscious—i.e., it

prints out "I am conscious"—and a good programmer can even program it so that it will carry on a rudimentary argument to the effect that it is conscious.
But that has nothing to do with whether or not it really is conscious.

As I discussed earlier, Searle frequently changes my statements in
critical ways, and in this case has left out the word "convincingly."
Of course one can trivially make a computer claim to be conscious. I
make the same point repeatedly. Claims to be conscious neither prove
nor even suggest its actual presence, nor does an inability to make
such a claim demonstrate a lack of consciousness. What I am assert-
ing, specifically, is that we will meet entities in several decades that
convincingly claim to be conscious.

Searle asserts that I assert that people will "just come to accept
without question that machines are conscious." This is a typical distortion of Searle. Many people will accept that machines are conscious precisely because the claims will be convincing. There is a
huge difference between idle claims (which are feasible today), and
convincing claims (which are not yet feasible). It is the difference
between the twentieth and twenty-first centuries, and one of the primary points of my book.

Now what does it mean to be convincing? It means that when a
nonbiological entity talks about its feelings, its behavior at that moment and subsequently will be fully consistent with what we would
expect of a human who professed such feelings. This requires enormously subtle, deep, and complex behavior. Nonbiological entities
today do not have this ability. What I am specifically claiming is that
twenty-first century nonbiological entities will.

This development will have enormous implications for the relationship between humans and the technology we will have created,
and I talk extensively in the book about these implications.

One of those implications is not that such entities are necessarily
conscious, even though their claims (to be conscious) will be convincing. We come back to the inability to penetrate the subjective

experience of another entity. We accept that other humans are conscious, but even this is a shared assumption. And humans are not of like mind when it comes to the consciousness of non-human entities such as animals. We can argue about the issue, but there is no definitive consciousness-detector that we can use to settle the argument. The issue of the potential consciousness of nonbiological entities will be even more contentious than the arguments we have today about the potential consciousness of non-human entities. My prediction is more a political prediction than a philosophical one.

As I mentioned earlier, Searle writes: "Actual human brains cause consciousness by a series of specific neurobiological processes in the brain."

Searle provides (and has provided) no basis for such a startling view. To illuminate where Searle is coming from, I take the liberty of quoting from a letter Searle sent me (dated December 15, 1998), in which he writes

> . . . it may turn out that rather simple organisms like termites or snails are conscious. . .The essential thing is to recognize that consciousness is a biological process like digestion, lactation, photosynthesis, or mitosis, and you should look for its specific biology as you look for the specific biology of these other processes.

I wrote Searle back:

> Yes, it is true that consciousness emerges from the biological process(es) of the brain and body, but there is at least one difference. If I ask the question, "does a particular entity emit carbon dioxide," I can answer that question through clear objective measurement. If I ask the question, "is this entity conscious," I may be able to provide inferential arguments—possibly strong and convincing ones—but not clear objective measurement.

With regard to the snail, I wrote:

> Now when you say that a snail may be conscious, I think what you are saying is the following: that we may discover a certain neurophysiological basis for consciousness (call it "x") in humans such that when this basis was present humans were conscious, and when it was not present humans were not conscious. So we would presumably have an objectively measurable basis for consciousness. And then if we found that in a snail, we could conclude that it was conscious. But this inferential conclusion is just a strong suggestion, it is not a proof of subjective experience on the snail's part. It may be that humans are conscious because they have "x" as well as some other quality that essentially all humans share, call this "y." The "y" may have to do with a human's level of complexity or something having to do with the way we are organized, or with the quantum properties of our tubules (although this may be part of "x"), or something else entirely. The snail has "x" but doesn't have "y" and so it may not be conscious.
>
> How would one settle such an argument? You obviously can't ask the snail. You can't tell from its fairly simple and more-or-less predictable behavior. Pointing out that it has "x" may be a good argument and many people may be convinced by it. But it's just an argument, it's not a direct measurement of the snail's subjective experience. Once again, objective measurement is incompatible with the very concept of subjective experience.
>
> And indeed we have such arguments today. Not about snails so much, but about higher level animals. It is

apparent to me that dogs and cats are conscious, and I
think you mentioned that you accept this as well. But
not all humans accept this. I can imagine scientific
ways of strengthening the argument by pointing out
many similarities between these animals and humans,
but again these are just arguments, not scientific proof.

Searle expects to find some clear biological "cause" of consciousness. And he seems unable to acknowledge that either understanding or consciousness may emerge from an overall pattern of activity. Other philosophers, such as Daniel Dennett, have articulated such "pattern emergent" theories of consciousness. But whether "caused" by a specific biological process or by a pattern of activity, Searle provides no foundation for how we would measure or detect consciousness. Finding a neurological correlate of consciousness in humans does not prove that consciousness is necessarily present in other entities with the same correlate, nor does it prove that the absence of such correlate indicates the absence of consciousness. Such inferential arguments necessarily stop short of direct measurement. In this way, consciousness differs from objectively measurable processes such as lactation and photosynthesis.

Searle writes in his chapter: "It is out of the question, for purely neurobiological reasons, to suppose that the chair or the computer is conscious."

Just what neurobiological reasons is Searle talking about? I agree that chairs don't seem to be conscious, but as for computers of the future that have the same complexity, depth, subtlety, and capabilities as humans, I don't think we can rule out the possibility that they are conscious. Searle just assumes that they are not, and that it is "out of the question" to suppose otherwise. There is really nothing more of a substantive nature to Searle's "arguments" than this tautology.

Now part of the appeal of Searle's stance against the possibility of a computer being conscious is that the computers we know today just don't seem to be conscious. Their behavior is brittle and formulaic, even if they are occasionally unpredictable. But as I pointed out above, computers today are still a million times simpler than the human brain, which is at least one reason they don't share all of the endearing qualities of human thought. But that disparity is rapidly shrinking, and will ultimately reverse itself in several decades. The twenty-first century machines I am talking about in the book will appear and act very differently than the relatively simple computers of today.

Searle may assert that the level of complexity and capacity is irrelevant, that even if nonbiological entities become trillions of times more complex and capable than humans, they inherently just don't have this mysterious neurobiological basis of consciousness. I have no problem with his believing that, but he should present it simply as his belief, and not wrap it in tautological arguments that provide no basis for such a belief.

The Chinese Room Argument is based on the idea that it seems ridiculous that a simple machine can be conscious. He then describes a simple machine successfully carrying out deep, extremely complex tasks such as answering unanticipated questions in Chinese. But simple machines would never accomplish such tasks. However, with regard to the extremely complex machines that could accomplish such difficult and subtle tasks, machines that would necessarily match or exceed the complexity of the human brain, the Chinese Room tells us nothing about their consciousness. It may be that consciousness emerges from certain types of very complex self-organizing processes that take place in the human brain. If so, then recreating the essence of these processes would also result in consciousness. It is certainly a plausible conjecture.

Searle writes:

> Kurzweil is aware of this objection and tries to meet it with a slippery-slope argument: We already have brain implants, such as cochlear implants in the auditory system, that can duplicate and not merely simulate certain brain functions. What is to prevent us from a gradual replacement of all the brain anatomy that

would preserve and not merely simulate our consciousness and the rest of our mental life? In answer
to this, I would point out that he is now abandoning
the main thesis of the book, which is that what is important for consciousness and other mental functions
is entirely a matter of computation. In his words, we
will become software, not hardware.

Once again, Searle misrepresents the essence of my argument.
As I described in my chapter in this book and in greater detail in my
book, I describe this slippery-slope scenario and then provide two
strong arguments: one that consciousness is preserved, and a second
argument that consciousness is not preserved. I present this specifically to illustrate the contradictions inherent in simplistic explanations of the phenomenon of consciousness. The difficulty of resolving this undeniably important issue, and the paradoxes inherent in
our understanding of consciousness, stem, once again, from our inability to penetrate subjective experience with objective measurement. I frequently present the perplexity of the issue of consciousness by showing how reasonable and logical arguments lead us to
contradictory conclusions. Searle takes one of these arguments completely out of context and then presents that as my position.

As for "abandoning the main thesis of [my] book, Searle's assertion is absurd. The primary thesis of the book is exactly this: that we
will recreate the processes in our brains, and then extend them, and
ultimately merge these enhanced processes into our human-machine
civilization. I maintain that these recreated nonbiological systems
will be highly intelligent, and use this term to refer to the highly
flexible skills that we exhibit as humans.

On the Difference Between
Intrinsic (Observer Independent) and
Observer-Relative Features of the World

With regard to Searle's argument that he numbers (2), I will respond briefly as many of the points have already been covered. First of all, I will point out that from a prevalent interpretation of quantum theory, all features of the world are rendered as observer-relative. But let's consider Searle's distinction as valid for the world as it appears to us.

Searle writes:

> In a psychological, observer-independent sense, I am more intelligent than my dog, because I can have certain sorts of mental processes that he cannot have, and I can use these mental capacities to solve problems that he cannot solve. But in this psychological sense of intelligence, wristwatches, pocket calculators, computers, and cars are not candidates for intelligence, because they have no mental life whatever.

Searle doesn't define what he means by mental life. But by any reasonable interpretation of the term, I would grant that Searle's observation is reasonable with respect to pocket calculators, cars, and the like. The statement is also reasonable with regard to today's computers. But as for the "computers" that we will meet a few decades from now, Searle's statement just reveals, once again, his bias that computers are inherently incapable of "mental life." It is an assumption that produces an identical conclusion, one of Searle's many tautologies.

If by "mental life," Searle is talking about our human ability to place ideas in a rich array of contexts, to deal with subjects in a fluid and subtle way, to recognize and respond appropriately to human emotions, and all of the other endearing and impressive qualities of our species, then computers (nonbiological entities) will achieve—according to the primary thesis of my book—these abilities and behaviors. If we're talking about consciousness, then we run into the same objective-subjective barrier.

Searle writes:

> In an observer-relative sense, we can indeed say that lots of machines are more intelligent than human beings because we have designed the machines in such a way as to help us solve problems that we cannot solve, or cannot solve as efficiently, in an unaided fashion. Chess-playing machines and pocket calculators are good examples. Is the chess-playing machine really more intelligent at chess than Kasparov? Is my pocket calculator more intelligent than I at arithmetic? Well, in an intrinsic or observer-independent sense, of course not, the machine has no intelligence whatever, it is just an electronic circuit that we have designed, and can ourselves operate, for certain purposes. But in the metaphorical or observer-relative sense, it is perfectly legitimate to say that the chess-playing machine has more intelligence, because it can produce better results. And the same can be said for the pocket calculator.

> There is nothing wrong with using the word "intelligence" in both senses, provided you understand the difference between the observer-relative and the observer-independent. The difficulty is that this word has been used as if it were a scientific term, with a scientifically precise meaning. Indeed, many of the exaggerated claims made on behalf of "artificial intelligence" have been based on this systematic confusion between observer-independent, psychologically relevant intelligence and metaphorical, observer-relative, psychologically irrelevant ascriptions of intelligence. There is nothing wrong with the metaphor as

such; the only mistake is to think that it is a scientifically precise and unambiguous term. A better term than "artificial intelligence" would have been "simulated cognition."

Exactly the same confusion comes over the notion of "computation." There is a literal sense in which human beings are computers because, for example, we can compute 2+2=4. But when we design a piece of machinery to carry out that computation, the computation 2+2=4 exists only relative to our assignment of a computational interpretation to the machine. Intrinsically, the machine is just an electronic circuit with very rapid changes between such things as voltage levels. The machine knows nothing about arithmetic just as it knows nothing about chess. And it knows nothing about computation either, because it knows nothing at all. We use the machinery to compute with, but that does not mean that the computation is intrinsic to the physics of the machinery. The computation is observer-relative, or to put it more traditionally, "in the eye of the beholder."

This distinction is fatal to Kurzweil's entire argument, because it rests on the assumption that the main thing humans do in their lives is compute. Hence, on his view, if—thanks to Moore's Law—we can create machines that can compute better than humans, we have equaled and surpassed humans in all that is distinctively human. But in fact humans do rather little that is literally computing. Very little of our time is spent working out algorithms to figure out answers to questions. Some brain processes can be usefully described as if they were computational, but that is observer-relative. That is like the attribution of compu-

tation to commercial machinery, in that it requires an
outside observer or interpreter.

There are many confusions in the lengthy quote above, several
of which I have already discussed. When I speak of the intelligence
that will emerge in twenty-first century machines as a result of reverse engineering the human brain and recreating and extending these
extensive processes in extremely powerful new substrates, I am not
talking about trivial forms of "intelligence" such as found in calculators and contemporary chess machines. I am not referring to the "narrow" victories of contemporary computers in areas such as chess,
diagnosing blood cell images, or tracking land terrain images in a
cruise missile. What I am talking about is recreating the processes
that take place in the human brain, which, as Searle acknowledges, is
a machine that follows natural laws in the physical world. It is disingenuous for Searle to maintain that I confuse the narrow calculations
of a calculator or even a game-playing algorithm with the sorts of
deep intelligence displayed by the human brain.

I do not maintain that the processes that take place in human
brains can be recreated in nonbiological machines because human
beings are capable of performing arithmetic. This is typical of Searle's
disingenuous arguments: attributing absurd assertions to my book
that in fact it never makes, and then pointing to their absurdity.

Another example is his false statement that I assume that the main
thing humans do in their lives is compute. I make the opposite point:
very little of our time is spent "computing." I make it clear that what
goes on in the human brain is a pattern recognition paradigm: the
complex, chaotic, and unpredictable interplay of millions of inter-
secting and interacting processes. We have in fact no direct means of
performing mental computation (in the sense that Searle refers to in
the above quote) at all. When we perform "computations" such as
figuring out 2+2, we use very indirect and complex means. There is
no direct calculator in our brains.

A nonbiological entity that contains an extended copy of the very extensive processes that take place in the human brain can combine the resulting human-like abilities with the speed, accuracy and sharing ability that constitute a current superiority of machines. As I mentioned above, humans are unable to directly transfer their knowledge to other persons. Computers, however, can share their knowledge very quickly. As we replicate the functionality of human neuron clusters, we are not leaving out quick downloading ports on the neurotransmitter strength patterns. Thus future machines will be able to combine human intellectual and creative strengths with machine strengths. When one machine learns a skill or gains an insight, it will be able to share that knowledge instantly with billions of other machines.

On the Church-Turing Thesis

Searle makes some strange statements about the Church-Turing Thesis, an important philosophical thesis independently presented by Alan Turing and Alonzo Church.

Searle writes:

> We know from the Church-Turing Thesis, a mathematical result, that any computation that can be carried out on a neural net can be carried out on a symbol-manipulating machine.

Searle also writes:

> [T]he basic idea [of the Church-Turing Thesis] is that any problem that has an algorithmic solution can be solved on a Turing machine, a machine that manipulates only two kinds of symbols, the famous zeroes and ones.

It is remarkable that Searle refers to the Church-Turing Thesis as a "mathematical result." He must be confusing the Church-Turing Thesis (CTT) with Church and Turing theorems. CTT is not a mathematical theorem at all, but rather a philosophical conjecture which relates to a proposed relationship between what a human brain can do and what a Turing Machine can do. There are a range of versions or interpretations of CTT. A standard version is that any method that a human can use to solve a mathematical problem in a finite amount of time can be expressed as a general recursive function and can therefore be solved in a finite amount of time on a Turing Machine. Searle's definition only makes sense if we interpret his phrase "algorithmic solution" to mean a method that a human follows, but that is not the common meaning of this phrase (unless we qualify the phrase as in "algorithmic solutions implemented by a human brain"). The phrase "algorithmic solution" usually refers to a method that can be implemented on a Turing Machine. This makes the Searle definition a tautology.

Broader versions of CTT consider problems beyond mathematical problems, which is consistent with the definition I offer in the book's timeline. The definition I provided is necessarily simplified as it is one brief entry in a lengthy timeline ("1937: Alonzo Church and Alan Turing independently develop the Church-Turing Thesis. This thesis states that all problems that a human being can solve can be reduced to a set of algorithms, supporting the idea that machine intelligence and human intelligence are essentially equivalent"). In this conception of CTT, I relate problems solvable by a human to algorithms, and use the word "algorithms" in its normal sense as referring to methods that can be implemented on a Turing Machine.

There are yet broader conceptions of CTT that relate the processes that take place in the human brain to methods that are computable. This conjecture is based on the following: (i) the constituent components of brains (e.g., neurons, interneuronal connections, synaptic clefts, neurotransmitter concentrations) are made up of matter and energy, therefore: (ii) these constituent components follow physical laws, therefore: (iii) the behavior of these components are describable in mathematical terms (even if including some irreducibly random elements), therefore: (iv) the behavior of such components is machine-computable.

In Conclusion

I believe that the scale of Searle's misrepresentation of ideas from the AI community stems from a basic lack of understanding of technology. He is stuck in a mindset that nonbiological entities are only capable of manipulating logical symbols, and appears to be unaware of other paradigms. It is true that manipulating symbols is largely how rule-based expert systems and game-playing programs such as Deep Blue work. But the current trend is in a different direction, towards self-organizing chaotic systems that employ biological-inspired methods, including processes derived directly from the reverse engineering of the hundreds of neuron clusters we call the human brain. Searle acknowledges that biological neurons are machines, indeed that the entire brain is a machine. Recent advances that I discussed above have shown that we can recreate in an extremely detailed way the "causal powers" of individual neurons as well as those of substantial neuron clusters. There is no conceptual barrier to scaling these efforts up to the entire human brain.

Searle argues that the Church-Turing Thesis (it's actually Church and Turing theorems) show that neural nets can be mapped onto algorithms that can be implemented in machines. Searle's own argument, however, can be applied equally well to *biological* neural nets, and indeed the experiments I cite above demonstrate this empirically.

Searle is a master of combining tautologies and contradictions in the same argument, but his illogical reasoning to the effect that machines that demonstrate understanding have no understanding does nothing to alter these rapidly accelerating developments.

7

Applying Organic Design Principles to Machines is Not an Analogy But a Sound Strategy: Response to Michael Denton

Ray Kurzweil

11.11.04

The Bridge is Already Under Construction

Similar to Dembski, Denton points out the apparent differences between the design principles of biological entities (e.g., people) and those of the machines he has known. Denton eloquently describes organisms as "self-organizing, . . . self-referential, . . . self-replicating, . . . reciprocal, . . . self-formative, and . . . holistic." He then makes the unsupported leap, a leap of faith one might say, that such organic forms can only be created through biological processes, and that such forms are "immutable, . . . impenetrable, and . . . fundamental" realities of existence.

I do share Denton's "awestruck" sense of "wonderment" at the beauty, intricacy, strangeness, and inter-relatedness of organic systems, ranging from the "eerie other-worldly impression" of asymmetric protein shapes to the extraordinary complexity of higher-order organs such as the human brain. Further, I agree with Denton that biological design represents a profound set of principles. However, it is precisely my thesis, which neither Denton nor the other critics represented in this book acknowledge nor respond to, that machines (i.e., entities derivative of human-directed design) can use—and already are using—these same principles. This has been the thrust of my own work, and in my view represents the wave of the future. Emulating the ideas of nature is the most effective way to harness the enormous powers that future technology will make available.

The concept of holistic design is not an either-or category, but rather a continuum. Biological systems are not completely holistic nor are contemporary machines completely modular. We can identify units of functionality in natural systems even at the molecular level, and discernible mechanisms of action are even more evident at the higher level of organs and brain regions. As I pointed out in my response to Searle, the process of understanding the functionality and information transformations performed in specific brain regions is well under way. It is misleading to suggest that every aspect of the human brain interacts with every other aspect, and that it is thereby impossible to understand its methods. Lloyd Watts, for example, has identified and modeled the transformations of auditory information in more than two dozen small regions of the human brain. Conversely, there are many examples of contemporary machines in which many of the design aspects are deeply interconnected and in which "bottom up" design is impossible. As one example of many, General Electric uses "genetic algorithms" to evolve the design of its jet engines as they have found it impossible to optimize the hundreds of deeply interacting variables in any other way.

Denton writes:

> Today almost all professional biologists have adopted
> the mechanistic/reductionist approach and assume that
> the basic parts of an organism (like the cogs of a watch)
> are the primary essential things, that a living organ-
> ism (like a watch) is no more than the sum of its parts,
> and that it is the parts that determine the properties of
> the whole and that (like a watch) a complete descrip-
> tion of all the properties of an organism may be had
> by characterizing its parts in isolation.

What Denton is ignoring here is the ability of complex processes to
exhibit *emergent* properties which go beyond "its parts in isolation."
Denton appears to recognize this potential in nature when he writes:
"In a very real sense organic forms . . . represent genuinely emergent
realities." However, it is hardly necessary to resort to Denton's "vitalistic model" to explain emergent realities. Emergent properties
derive from the power of patterns, and there is nothing that restricts
patterns and their emergent properties to natural systems.

Denton appears to acknowledge the feasibility of emulating the
ways of nature, when he writes:

> Success in engineering new organic forms from pro-
> teins up to organisms will therefore require a com-
> pletely novel approach, a sort of designing from 'the
> top down.' Because the parts of organic wholes only
> exist in the whole, organic wholes cannot be speci-
> fied bit by bit and built up from a set of relatively
> independent modules; consequently the entire undi-
> vided unity must be specified together *in toto*.

Here Denton provides sound advice and describes an approach
to engineering that I and other researchers use routinely in the areas
of pattern recognition, complexity (also called chaos) theory, and

self-organizing systems. Denton appears to be unaware of these methodologies and after describing examples of bottom-up component-driven engineering and their limitations concludes with no justification that there is an unbridgeable chasm between the two design philosophies. The bridge is already under construction.

How to Create Your Own "Eerie Other-Worldly" But Effective Designs: Applied Evolution

In my book I describe how to apply the principles of evolution to creating intelligent designs. It is an effective methodology for problems that contain too many intricately interacting aspects to design using the conventional modular approach. We can, for example, create (in the computer) millions of competing designs, each with their own "genetic" code. The genetic code for each of these design "organisms" describes a potential solution to the problem. Applying the genetic method, these software-based organisms are set up to compete with each other and the most successful are allowed to survive and to procreate. "Offspring" software entities are created, each of which inherits the genetic code (i.e., the design parameters) of two parents. Mutations and other "environmental challenges" are also introduced. After thousands of generations of such simulated evolution, these genetic algorithms often produce complex original designs. In my own experience with this approach, the results produced by genetic algorithms are well described by Denton's description of organic molecules in the "apparent illogic of the design and the lack of any obvious modularity or regularity...the sheer chaos of the arrangement...[and the] almost eerie other-worldly non-mechanical impression."

Genetic algorithms and other *top-down* self-organizing design methodologies (e.g., neural nets, Markov models) incorporate an unpredictable element, so that the results of such systems are actually different every time the process is run. Despite the common wisdom that machines are deterministic and therefore predictable, there are numerous readily available sources of randomness available to machines. Contemporary theories of quantum mechanics postulate profound quantum randomness at the core of existence. According to quantum theory, what appears to be the deterministic behavior of systems at a macro level is simply the result of overwhelming statistical preponderancies based on enormous numbers of fundamentally unpredictable events. Moreover, the work of Stephen Wolfram and others has demonstrated that even a system that is in theory fully deterministic can nonetheless produce effectively random results.

The results of genetic algorithms and similar "self-organizing" approaches create designs which could not have been designed through a modular component-driven approach. The "strangeness... chaos, ... the dynamic interaction" of parts to the whole that Denton attributes only to organic structures describe very well the qualities of the results of these human initiated chaotic processes.

In my own work with genetic algorithms, I have examined the process in which a genetic algorithm gradually improves a design. It accomplishes this precisely through an incremental "all at once" approach, making many small, distributed changes throughout the design which progressively improve the overall fit or "power" of the solution. A genetic algorithm does not accomplish its design achievements through designing individual subsystems one at a time. The entire solution emerges gradually, and unfolds from simplicity to complexity. The solutions it produces are often asymmetric and ungainly, but effective, just as in nature. Often, the solutions appear elegant and even beautiful.

Denton is certainly correct that most contemporary machines are designed using the modular approach. It is important to note that there are certain significant engineering advantages to this traditional approach to creating technology. For example, computers have far more prodigious and accurate memories than humans, and can perform certain types of transformations far more effectively than unaided human intelligence. Most importantly, computers can share their memories and patterns instantly. The chaotic non-modular approach also has clear advantages which Denton well articulates, as evidenced by the deep prodigious powers of human pattern recognition. But it is a wholly unjustified leap to say that because of the current (and diminishing!) limitations of human-directed technology that biological systems are inherently, even ontologically, a world apart. The exquisite designs of nature have benefited from a profound evolutionary process. Our most complex genetic algorithms today incorporate genetic codes of thousands of bits whereas biological entities such as humans are characterized by genetic codes of billions of bits (although it appears that as a result of massive redundancies and other inefficiencies, only a few percent of our genome is actually utilized). However, as is the case with all information-based technology, the complexity of human-directed evolutionary engineering is increasing exponentially. If we examine the rate at which the complexity of genetic algorithms and other nature-inspired methodologies are increasing, we find that they will match the complexity of human intelligence within a few decades.

To Fold a Protein

Denton points out we have not yet succeeded in folding proteins in three dimensions, "even one consisting of only 100 components." It should be pointed out, however, that it is only in the recent few years that we have had the tools even to visualize these three-dimensional patterns. Moreover, modeling the interatomic forces will require on the order of a million billion calculations per second, which is beyond the capacity of even the largest supercomputers available today. But computers with this capacity are expected soon. IBM's "Blue Gene" computer, scheduled for operation in 2005, will have precisely this capacity, and as the name of the project suggests, is targeted at the protein-folding task.

We have already succeeded in cutting, splicing, and rearranging genetic codes, and harnessing nature's own biochemical factories to produce enzymes and other complex biological substances. It is true that most contemporary work of this type is done in two dimensions, but the requisite computational resources to visualize and model the far more complex three-dimensional patterns found in nature is not far from realization.

In discussing the prospects for solving the protein-folding problem with Denton himself, he acknowledged that the problem would eventually be solved, estimating that it was perhaps a decade away. The fact that a certain technical feat has not *yet* been accomplished is not a strong argument that it never will.

Denton writes:

> From knowledge of the genes of an organism it is impossible to predict the encoded organic forms. Neither the properties nor structure of individual proteins nor those of any higher order forms—such as ribosomes and whole cells—can be inferred even from the most exhaustive analysis of the genes and their primary products, linear sequences of amino acids.

Although Denton's observation above is essentially correct, this only points out that the genome is only part of the overall system. The DNA code is not the whole story, and the rest of the molecular support system is needed for the system to work and for it to be understood.

I should also point out that my thesis on recreating the massively parallel, digitally controlled analog, hologram-like, self-organizing and chaotic processes of the human brain does not require us to fold proteins. There are dozens of contemporary projects which have succeeded in creating detailed recreations of neurological systems, including neural implants which successfully function inside people's brains, without folding any proteins. However, I understand Denton's argument about proteins to be an essay on the holistic ways of nature. But as I have pointed out, there are no essential barriers to our emulating these ways in our technology, and we are already well down this path.

But b/c we emulate them, does that mean they are the same thing.

Contemporary Analogues to Self-Replication

Denton writes:

> To begin with, every living system replicates itself,
> yet no machine possesses this capacity even to the
> slightest degree....Living things possess the ability
> to change themselves from one form into
> another...The ability of living things to replicate them-
> selves and change their form and structure are truly
> remarkable abilities. To grasp just how fantastic they
> are and just how far they transcend anything in the
> realm of the mechanical, imagine our artifacts en-
> dowed with the ability to copy themselves and—to
> borrow a term from science fiction—"morph" them-
> selves into different forms.

First of all, we do have a new form of self-replicating entity that
is human-made, and which did not exist a short while ago, namely
the computer (or software) virus. Just as biological self-replicating
entities require a medium in which to reproduce, viruses require the
medium of computers and the network of networks known as the
Internet. As far as changing form is concerned, some of the more
advanced and recent software viruses demonstrate this characteristic. Moreover, morphing form is precisely what happens in the case
of the reproducing designs created by genetic algorithms. Whereas
most software viruses reproduce asexually, the form of self-replication harnessed in most genetic algorithms is sexual (i.e., utilizing
two "parents" such that each offspring inherits a portion of its genetic code from each parent). If the conditions are right, these evolving software artifacts do morph themselves into different forms, indeed into increasingly complex forms that provide increasingly greater
power in solving nontrivial problems. And lest anyone think that there
is an inherent difference between these evolving software entities
and actual physical entities, it should be pointed out that software

entities created through genetic algorithms often do represent the designs of physical entities such as engines or even of robots, as recently demonstrated by scientists at Tufts. Conversely, biological physical entities such as humans are also characterized by the data contained in their genetic codes.

Nanobots, which I described in the first chapter of this book, will also provide the ability to create morphing structures in the physical world. J.D. Storrs, for example, has provided designs of special nanobots, which he calls "foglets," which will eventually be capable of organizing and reorganizing themselves into any type of physical structure, thereby bringing the morphing qualities of virtual reality into real reality.

On Consciousness and the Thinking Ability of Humans

Denton writes:

> Finally I think it would be acknowledged by even ardent advocates of strong AI like Kurzweil, Dennett and Hofstadter that *no machine has been built to date which exhibits consciousness and can equal the thinking ability of humans.* Kurzweil himself concedes this much in his book. As he confesses: "Machines today are still a million times simpler than the human brain. . . . Of course Kurzweil believes, along with the other advocates of strong AI that sometime in the next century computers capable of carrying out 20 million billion calculations per second (the capacity of the human brain) will be achieved and indeed surpassed. And in keeping with the mechanistic assumption that organic systems are essentially the same as machines then of course such machines will equal or surpass the intelligence of man. . . . Although the mechanistic faith in the possibility of strong AI still runs strong

among researchers in this field, Kurzweil being no exception, there is no doubt that *no one has manufactured anything that exhibits intelligence remotely resembling that of man.*

First of all, my positions are neither concessions nor confessions. Our technology today is essentially where I had expected it to be by this time when I wrote a book (*The Age of Intelligent Machines*) describing the law of accelerating returns in the 1980s. Once again, Denton's accurate observation about the limitations of today's machines is not a compelling argument on inherent restrictions that can never be overcome. Denton himself acknowledges the quickening pace of technology that is moving "at an ever-accelerating rate one technological advance [following] another."

Denton is also oversimplifying my argument in the same way that Searle does. It is not my position that once we have computers with a computing capacity comparable to that of the human brain, that "of course such machines will equal or surpass the intelligence of man." I state explicitly in the first chapter of this book and in many different ways in my book *The Age of Spiritual Machines* that "this level of processing power is a necessary but not sufficient condition for achieving human-level intelligence in a machine." The bulk of my thesis addresses the issue of how the combined power of exponentially increasing computation, communication, miniaturization, brain scanning, and other accelerating technology capabilities, will enable us to reverse engineer, that is to understand, and then to recreate in other forms, the methods underlying human intelligence.

Finally, Denton appears to be equating the issue of "*exhibit[ing] consciousness*" with the issue of "*equal[ing] the thinking ability of humans.*" Without repeating the arguments I presented in both the first chapter of this book and in my response to Searle, I will say that these issues are quite distinct. The latter issue represents a salient goal of *objective* capability, whereas the former issue represents the essence of *subjective* experience.

In summary, Denton is far too quick to conclude that complex systems of matter and energy in the physical world are incapable of exhibiting the "emergent . . . vital characteristics of organisms such as self-replication, "morphing," self-regeneration, self-assembly and the holistic order of biological design," and that, therefore, "organisms and machines belong to different categories of being." Dembski and Denton share the same limited view of machines as entities that can only be designed and constructed in a modular way. We can build (and already are building) "machines" that have powers far greater than the sum of their parts by combining the chaotic self-organizing design principles of the natural world with the accelerating powers of our human-initiated technology. The ultimate result will be a formidable combination indeed.

8

Dembski's Outdated Understanding: Response to William Dembski

Ray Kurzweil

12.2.04

Intelligence versus Consciousness

I cannot resist starting my response with an amusing misquotation. Dembski writes: "Those humans who refuse to upload themselves will be left in the dust, becoming 'pets,' as Kurzweil puts it, of the newly evolved computer intelligences." This is indeed a quotation from my book. But the reference Dembski attributes to me is actually from Ted Kaczynski (the "Unabomber"), not someone whose views I concur with. I'm sure it's an honest mistake, but a good example of how inattentive reading often results in people seeing only what they expect to see. That having been said, I will say that the misquotations from Dembski are not nearly on the massive scale of Searle.

Dembski is correct that with regard to human *performance*, indeed with regard to any of our objectively observed abilities and reactions, it is my view that what Dembski calls the materialist approach is valid. One might call this "capability materialism." Capability materialism is based on the observation that biological neurons and their interconnections are made up of matter and energy, that their methods can be described, understood, and modeled with either replicas or functionally equivalent recreations. As I pointed out at length earlier, we are already recreating functionally equivalent recreations of substantial neuron clusters, and there are no fundamental barriers to extending this process to the several hundred neural regions we call the human brain. I use the word "capability" because this includes all of the rich, subtle, and diverse ways in which humans interact with the world, not just those narrower skills that one might label as intellectual. Indeed, our ability to understand and respond to emotions is more complex and diverse than our ability to process intellectual issues.

Searle, for example, acknowledges that human neurons are biological machines. Few serious observers have postulated capabilities or reactions of human neurons that require Dembski's "extra-material factors." In my view, relying on the patterns of matter and energy in the human body and brain to explain its behavior and proficiencies need not diminish our wonderment at its remarkable qualities. Dembski has an outdated understanding of the concept of "machine," as I will detail below.

However, with regard to the issue of *consciousness*, I would have to say that Dembski and I are in agreement, although Dembski apparently does not realize this. He writes:

> The great mistake in trying to understand the mind-body problem is to suppose that it is a scientific problem. It is not. It is a problem of ontology (i.e., that branch of metaphysics concerned with what exists).

If by the "mind-body problem," Dembski means the issue of consciousness, then I agree with Dembski's statement. As I explained in my first chapter in this book and in my response to Searle, there is no objective (i.e., scientific) method that can definitively measure or determine the subjective experience (i.e., the consciousness) of another entity. We can measure correlates of subjective experience (e.g., outward or inward behavior, i.e., patterns of neuron activity), and we can use these correlates to make arguments about the potential consciousness of another entity (such as an animal or a machine), but these arguments remain just that. Such observations do not constitute objective proof of another entity's subjective experiences, i.e., of its consciousness. It comes down to the essential difference between the concepts of "objective" and "subjective."

As I pointed out, however, with multiple quotations of John Searle (e.g., "human brains cause consciousness by a series of specific neurobiological processes in the brain"), Searle apparently does believe that the essential philosophical issue of consciousness is determined by what Dembski calls "tender-minded materialism."

The arguments of scientist-philosophers such as Roger Penrose that consciousness in the human brain is somehow linked to quantum computing does not change the equation because quantum effects are properly part of the material world. Moreover there is nothing that prevents our utilizing quantum effects in our machines. Indeed, we are already doing this. The conventional transistor relies on the quantum effect of electron tunneling.

So the line-up on these issues is not as straightforward as might at first appear.

Dembski's Limited Understanding of Machines and Emergent Patterns

Dembski writes:

> [P]redictability is materialism's main virtue... We long for freedom, immortality, and the beatific vision... The problem for the materialist, however, is that these aspirations cannot be redeemed in the coin of matter.

> Unlike brains, computers are neat and precise . . . computers operate deterministically.

These and other statements of Dembski reveal a view of machines, or entities made up of patterns of matter and energy (i.e., "material" entities), that is limited to the literally simple-minded machines of nineteenth century automata. These machines with their hundreds, maybe thousands of parts were quite predictable and certainly not capable of longings for freedom and other such endearing qualities of the human entity. The same observations largely hold true for today's machines with their billions of parts. But the same cannot necessarily be said for machines with *millions of billions* of interacting "parts," entities with the complexity of the human brain and body.

First of all, it is incorrect to say that materialism is predictable. Even today's computer programs routinely use simulated randomness. If one needs truly random events in a process, there are devices that can provide this as well. Fundamentally, everything we perceive in the material world is the result of many trillions of quantum events, each of which display profound and irreducible quantum randomness at the core of physical reality. The material world—at both the macro and micro levels—is anything but predictable.

Although many computer programs do operate the way Dembski describes, the predominant methods in my own field of pattern recognition use biological-inspired methods called "chaotic computing," in which the unpredictable interaction of millions of processes, many of which contain random and unpredictable elements, provide unexpected yet appropriate answers to subtle questions of recognition. It is also important to point out that the bulk of human intelligence consists of just these sorts of pattern recognition processes.

As for our responses to emotions and our highest aspirations, these are properly regarded as emergent properties, profound ones to be sure, but nonetheless emergent patterns that result from the interaction of the human brain with its complex environment. The complexity and capacity of nonbiological entities is increasing exponentially and will match biological systems including the human brain (along with the rest of the nervous system and the endocrine system) within three decades. Indeed many of the designs of future machines will be biologically inspired, that is to say derivative of biological designs (this is already true of many contemporary systems). It is my thesis that by sharing the complexity as well as the actual patterns of human brains, these future nonbiological entities will display the intelligence and emotionally rich reactions of humans. They will have aspirations because they will share these complex emergent patterns.

Will such nonbiological entities be conscious? Searle claims that we can (at least in theory) readily resolve this question by ascertaining if it has the correct "specific neurobiological processes." It is my view that many humans, ultimately the vast majority of humans, will come to believe that such human-derived but nonetheless nonbiological intelligent entities are conscious, but that's a political prediction, not a scientific or philosophical judgement. Bottom line, I agree with Dembski that this is not a scientific question. Some observers go on to say that if it's not a scientific question, then it's not an important or even a real question. My view (and I'm sure Dembski agrees) is that because the question is not scientific, it is precisely for that reason a philosophical one, indeed the fundamental philosophical question.

Transcendence, Spirituality and God

Dembski writes:

> We need to transcend ourselves to find ourselves. Now the motions and modifications of matter offer no opportunity for transcending ourselves. . . . Freud . . . Marx . . . Nietzsche . . . each regarded the hope for transcendence as a delusion.

Dembski's view of transcendence as an ultimate goal is reasonably put. But I disagree that the material world offers no "opportunity for transcending." The material world inherently evolves, and evolution represents transcendence. As I wrote in the first chapter in this book, "Evolution moves towards greater complexity, greater elegance, greater knowledge, greater intelligence, greater beauty, greater creativity, greater love. And God has been called all these things, only without any limitation: infinite knowledge, infinite intelligence, infinite beauty, infinite creativity, and infinite love. Evolution does not achieve an infinite level, but as it explodes exponentially, it certainly moves in that direction. So evolution moves inexorably towards our conception of God, albeit never reaching this ideal."

Dembski writes:

> [A] machine is fully determined by the constitution, dynamics, and interrelationships of its physical parts. . . . "[M]achines" stresses the strict absence of extra-material factors. . . . The replacement principle is relevant to this discussion because it implies that machines have no substantive history. . . . But a machine, properly speaking, has no history. Its history is a superfluous rider—an addendum that could easily have been different without altering the machine. . . . For a machine, all that is, is what it is at this moment. . . . Machines access or fail to access items in storage. . . Mutatis mutandis, items that represent counterfactual occurrences (i.e., things that never happened) but which are accessible can be, as far as the machine is concerned, just a though they did happen.

It is important to point out that the whole point of my book and the first chapter of this book is that many of our dearly held assumptions about the nature of machines and indeed of our own human nature will be called into question in the next several decades. Dembski's conception of "history" is just another aspect of our humanity that necessarily derives from the richness, depth and complexity of being human. Conversely, not having a history in the Dembski sense is just another attribute of the simplicity of the machines that we have known up to this time. It is precisely my thesis that machines of the mid to late twenty-first century will be of such great complexity and richness of organization that their behavior will evidence emotional reactions, aspirations, and, yes, history. So Dembski is merely describing today's limited machines and just assuming that these limitations are inherent. This line of argument is entirely equivalent to stating that "today's machines are not as capable as humans, therefore machines will never reach this level of performance." Dembski is just assuming his conclusion. *is he?*

Dembski's view of the ability of machines to understand their own history is limited to "accessing" items in storage. But future machines will possess not only a record of their own history, but an ability to understand that history and to reflect insightfully upon it. As for "items that represent counterfactual occurrences," surely the same can be said for our human memories.

Dembski's lengthy discussion of spirituality is summed up by his closing paragraph of his "Humans as Spiritual Machines" section:

> But how can a machine be aware of God's presence? Recall that machines are entirely defined by the constitution, dynamics, and interrelationships among their physical parts. It follows that God cannot make his presence known to a machine by acting upon it and thereby changing its state. Indeed, the moment God acts upon a machine to change its state, it no longer properly is a machine, for an aspect of the machine now transcends its physical constituents. It follows that awareness of God's presence by a machine must be independent of any action by God to change the state of the machine. How then does the machine come to awareness of God's presence? The awareness must

be self-induced. Machine spirituality is the spirituality of self-realization, not the spirituality of an active God who freely gives himself in self-revelation and thereby transforms the beings with which he is in communion. For Kurzweil to modify "machine" with the adjective "spiritual" therefore entails an impoverished view of spirituality.

Dembski states that an entity (e.g., a person) cannot be aware of God's presence without God acting upon her, yet God cannot act upon a machine, so therefore a machine cannot be aware of God's presence. This reasoning here is entirely tautological and human-centric. God only communes with humans, and only biological ones at that. I have no problem with Dembski believing this as a personal belief, but he fails to the make the "strong case" that he promises that "humans are not machines—period." As with Searle, Dembski just assumes his conclusion.

Where Can I Get Some of Dembski's "Extra-Material" Thinking Stuff?

Like Searle, Dembski cannot seem to grasp the concept of the emergent properties of complex distributed patterns. He writes:

> Anger presumably is correlated with certain localized brain excitations. But localized brain excitations hardly explain anger any better than overt behaviors associated with anger, like shouting obscenities. Localized brain excitations may be reliably correlated with anger, but what accounts for one person interpreting a comment as an insult and experiencing anger, and another person interpreting that same comment as a joke and experiencing laughter? A full materialist account of mind needs to understand localized brain excitations in terms of other localized brain

excitations. Instead we find localized brain excitations (representing, say, anger) having to be explained in terms of semantic contents (representing, say, insults). But this mixture of brain excitations and semantic contents hardly constitutes a materialist account of mind or intelligent agency.

Dembski assumes that anger is correlated with a "localized brain excitation," but anger is almost certainly the reflection of complex distributed patterns of activity in the brain. Even if there is a local- ized neural correlate associated with anger, it nonetheless results from multifaceted and interacting patterns. Dembski's question as to why different people react differently to similar situations hardly requires us to resort to his extra-material factors for an explanation. The brains and experiences of different people are clearly not the same and these differences are well explained by differences in our physical brains.

It is useful to consider the analogy of the brain's organization to a hologram (a piece of film containing an interference pattern created by the interaction between a three-dimensional image and a laser light). When one looks through a hologram, one sees the original three-dimensional image, but none of the features of the image can be seen directly in the apparently random patterns of dots that are visible if one looks directly at the piece of film. So where are the features of the projected image? The answer is that each visual feature of the projected image is distributed throughout the entire pattern of dots that the hologram contains. Indeed, if you tear a hologram in half (or even in a large number of pieces), each piece will contain the entire image (albeit at reduced resolution). The visible image is an emergent property of the hologram's distributed pattern, and none of the image's features can be found through a localized analysis of the information in the hologram. So it is with the brain.

It is also the case that the human brain has a great deal of redundancy and it contains far more neural circuitry than is minimally needed to performs its functions. It is well known that the left and right halves of the brain, while not identical, are each sufficient to provide a more-or-less normal level of human functioning, which explains Louis Pasteur's intellectual accomplishments after his cerebral accident. Half a brain is enough.

I find it remarkable that Dembski cites the case of John Lorber's reportedly brainless patient as evidence that human intellectual functioning is the result of "extra-material factors." First of all, we need to take this strange report with a grain of salt. Many commentators have pointed out that Lorber's conclusion that his patient's brain was only 1 millimeter thick was flawed. As just one of many such critics, neurosurgeon Kenneth Till commented on the case of Lorber's patient: "Interpreting brain scans can be very tricky. There can be a great deal more brain tissue in the cranium than is immediately apparent."

It may be true that this patient's brain was smaller than normal, but that would not necessarily be reflected in obviously degraded capabilities. In commenting on the Lorber case, University of Indiana Professor Paul Pietsch writes, "How could this [the Lorber case] possibly be? If the way the brain functions is similar to the way a hologram functions, that [diminished brain size] might suffice. Certain holograms can be smashed to bits, and each remaining piece can reproduce the whole message. A tiny fragment of this page, in contrast, tells little about the whole story."

Even Lorber himself does not resort to "extra-material factors" to explain his observations. Lorber concludes that "there must be a tremendous amount of redundancy or spare capacity in the brain, just as there is with kidney and liver." Few commentators on this case resort to Dembski's "extra-material factors" to explain it.

Dembski's resolution of the ontology problem is to say that the ultimate basis of what exists is the "real world of things," things irreducible to material stuff. Dembski does not list what "things" we might consider as fundamental but presumably human minds would be on the list, and perhaps other "things" such as money and chairs. There may be a small congruence of our views in this regard. I regard Demski's things as patterns. Money, for example, is a vast and persisting pattern of agreements, understandings, and expectations.

"Ray Kurzweil" is perhaps not so vast a pattern, but thus far is also persisting. Dembski apparently regards patterns as ephemeral and not substantial, but as a pattern recognition scientist, I have a profound respect for the power and endurance of patterns. It is not unreasonable to regard patterns as a fundamental ontological reality. We are unable to really "touch" matter and energy directly, but we do directly experience the patterns underlying "things." *This is the issue*

Fundamental to my thesis is that as we apply our intelligence and the extension of our intelligence called technology to understanding the powerful patterns in our world (e.g., human intelligence), we can recreate—and extend!—these patterns in other substrates (i.e., with other materials). The patterns are more important than the materials that embody them.

Finally, if Dembski's intelligence-enhancing extra-material stuff really exists, then I'd like to know where I can get some.

9

What Turing Fallacy?
Response to Thomas Ray

Ray Kurzweil

R.L. 24

Measurement Without Observation

In Thomas Ray's articulated world, there is no such thing as consciousness, a view he makes clear in his reductionist view of quantum mechanics. Ray states that "it is the act of measurement that causes the collapse of the wave function, not conscious observation of the measurement." In other words, according to Ray, the collapse of the wave function is caused by measurement without observation, but what could this refer to? We know that any quantum "collapsed" event involving one or more particles causes some reaction beyond those particles immediately involved. No particle is an "island," so to speak. These inevitable reactions are properly considered measurements. The only conceivable circumstance in which an event would not cause a specific reaction (i.e., a measurement) is if

the event was indeterminate because the wave function was not collapsed. It is only through the collapse of the wave function that an
event becomes determinate and thereby causes a reaction in the world,
which constitutes a measurement. It is, therefore, the collapse of the
wave function that causes measurement, which Ray tells us causes
collapse of the wave function. So what Ray is saying is that the collapse of the wave function causes the collapse of the wave function.
By removing the concept of observation from measurement, Ray's
explanation of quantum mechanics devolves into this tautology.

Ray goes on to call any other view, even those of other main-
stream scientists, a "glaring...error." Ray's rigid view regards any
introduction of consciousness in the world to be an "error." I would
also point out that if we accept Ray's view, then Penrose's objection
to the potential of consciousness in a nonbiological entity (i.e.,
Penrose's argument to the effect that such an entity would have to
recreate the precise quantum state of a biological conscious entity)
becomes even less valid.

Colloquial Chaos Indeed

Ray casts doubt that there is increased chaos as an organism moves
from conception as a single cell to a mature individual. Consider
what we know about this process. The human genome contains 3
billion DNA rungs for a total of 6 billion bits of data. There is enormous redundancy in this information (e.g., a sequence known as
"ALU" is repeated hundreds of thousands of times), so the amount
of unique information is estimated at around 3% or about 23 mega-
bytes. In contrast, the human brain contains on the order of 100 trillion connections. Just specifying this connection data would require
trillions of bytes. Thus as we go from the genome, which specifies
the brain among all other organs, to the fully expressed individual,
the amount of information, considering just the brain connection patterns alone, increases by a factor of millions. We know that the genome specifies a wiring plan for the interneuronal connections that
includes a great deal of randomness, i.e., chaos, at specific imple-

mentation stages. This includes the stage of fetal wiring, during which interneuronal connections essentially wire themselves with a significant element of randomness applied during the process, as well as the growing of new dendrites after birth (which is believed to be a critical part of the learning process). This is at least one source of the increasing chaos resulting from the development of the individual from a fertilized egg. Another source is the chaos inherent in the environment that the individual encounters.

Ray states that I argue that "increasing rates of mutations and unpredictable events are, in part, driving the increasing frequency of 'salient events' in evolution." That's not my argument at all and I never make this statement. My position is that the acceleration of an evolutionary process, including both biological and technological evolution, results from the greater power of each level of evolution to create the next level. For example, with the evolution of DNA, evolutionary experiments could proceed more rapidly and more effectively with each stage of results recorded in the evolving DNA code. With the innovation of sexual reproduction, a more effective means for devising new combinations of genetic information became available. Within technological evolution, we also find that each generation of technology enables the next generation to proceed more rapidly. For example, the first generation of computers were designed by pen on paper and built with screwdrivers and wires. Compare that to the very rapid design of new computers using today's computer-assisted design tools.

Ray's Digital Fallacy

Thomas Ray starts his chapter by citing my alleged "failure to consider the unique nature of the digital medium." However, my thesis repeatedly refers to combining analog and digital methods in the same way that the human brain does; for example, "more advanced neural nets [which] are already using highly detailed models of human neurons, including detailed nonlinear analog activation functions." I go on to cite the advantages of emulating the brain's "digital controlled analog" design, and conclude that "there is a significant efficiency advantage to emulating the brain's analog methods." Analog methods are not the exclusive province of biological systems. We used to refer to "digital computers" to distinguish them from the more ubiquitous analog computers which were widely used during World War II.

It is also worth pointing out that analog processes can be emulated with digital methods whereas the reverse is not necessarily the case. However, there are efficiency advantages to analog processing as I point out above. Analog methods are readily recreated by conventional transistors which are essentially analog devices. It is only by adding the additional mechanism of comparing the transistor's output to a threshold that it is made into a digital device.

What Turing Fallacy?

Thomas Ray states:

> The primary criticism that I wish to make of Kurzweil's book, is that he proposes to create intelligent machines by copying human brains into computers. We might call this the Turing Fallacy. The Turing Test suggests that we can know that machines have become intelligent when we cannot distinguish them from human, in free conversation over a teletype. The Turing Test is one of the biggest red-herrings in science.

This paragraph contains several ideas, unrelated ones in my view, but concepts that appear to animate Ray's discomfort with "strong AI." I should make the caveat that Ray appears to accept the possibility of an advanced but "fundamentally alien intelligence" that is "rooted in and natural to the medium," but he dismisses the possibility, as well as the desirability, of AIs sharing human-like attributes. He states that "AIs must certainly be non-Turing," which he defines as "unlike human intelligences." So this is what Ray means by the "Turing Fallacy." He is maintaining that any intelligence that might emerge in nonbiological mediums would not and could not be a human intelligence, and would, therefore, be unable to pass the Turing Test.

It is fair to conclude from this that Thomas Ray accepts the Turing Test as a reasonable test for "human intelligence," but makes the point that there could be an alien intelligence that is very capable in terms of performing intelligent tasks but unable to pass this particular test. I should point out that Turing has made precisely the same point. Turing intended his Test specifically as a measure of human intelligence. An entity passing the Turing Test may be said to be intelligent, and moreover, to possess a human-form of intelligence. Turing specifically states that the converse statement does not hold, that failure to pass the Test does not indicate a lack of intelligence. I've made the same point. As I stated in the *Age of Intelligent Machines*, certain animals such as dolphins, giant squids, and certain species of whales, appear to have relatively high levels of intelligence, but are in no position to pass the Turing Test (they can't type for one thing). Even a human would not be able to pass the Test if she didn't speak the language of the Turing Test judge. The key point of the Turing test is that human language is sufficiently broad that we can test for the full range of human intelligence through human language dialogues. The Turing Test itself does not represent a fallacy, but rather a keen insight into the power of human communication through language to represent our thinking processes.

So where is the fallacy? Ray appears to be objecting to the concept of creating a human-like intelligence, one whose communication is sufficiently indistinguishable from human language-based communication such that it could pass the Turing Test, on grounds of both desirability and feasibility.

With regard to the first issue, the desirability of machines understanding our language is, I believe, clear from the entire history of the AI field. As machines have gained proficiency in aspects of human language, they have become more useful and more valuable.

Language is our primary method of communication, and machines need to understand human language in order to interact with us in an efficacious manner. Ultimately, they will need to understand human knowledge in a deep way to fully manifest their potential to be our partners in the creation of new knowledge. And as Turing pointed out, understanding human knowledge, including our capacity for understanding and expressing higher order emotions, is a prerequisite for effective human communication, and is therefore, a necessary set of skills to pass the Turing Test.

With regard to the issue of feasibility, Thomas Ray states:

> I accept that this level of computing power is likely to be reached, someday. But no amount of raw computer power will be intelligent in the relevant sense unless it is properly organized. This is a software problem, not a hardware problem.

I agree with this, of course, as I've had to state several times. A primary scenario that I describe in solving the "software problem" is to reverse engineer the methods deployed by the human brain. Although I also talk about the potential to copy specific human brains, I acknowledge that this is a more difficult task and will take longer. Reverse engineering the human brain is quite feasible, and we are further along in this project than most people realize. As I pointed out earlier, there are many contemporary examples that have demonstrated the feasibility of reverse engineering human neurons, neuron clusters, and entire brain regions, and then implementing the resulting detailed mathematical models in nonbiological mediums.

I mentioned Lloyd Watts' work in which he has developed a detailed and working model of more than a dozen brain regions related to auditory processing. Carver Mead's retina models, which are implemented as digital controlled analog processes on silicon chips, capture processes similar to those that take place in human visual processing. The complexity and level of detail in these models is expanding exponentially along with the growing capacity of our computational and communication technologies. This undertaking is similar to the genome project in which we scanned the genome and are now proceeding to understand the three-dimensional structures and processes described therein. In the mission to scan and reverse engineer the neural organization and information processing of the human brain, we are approximately where we were in the genome project about ten years ago. I estimate that we will complete this project within thirty years, which takes into account an exponential rather than linear projection of our anticipated progress.

Thomas Ray essentially anticipates the answer to his own challenge, when he states that:

> In order for the metallic "copy" to have the same function, we would have to abstract the functional properties out of the organic neural elements, and find structures and processes in the new metallic medium that provide identical functions. This abstraction and functional-structural translation from the organic into the metallic medium would require a deep understanding of the natural neural processes, combined with the invention of many computing devices and processes which do not yet exist.

I'm not sure why Ray uses the word "metallic" repeatedly, other than to demonstrate his inclination to regard nonbiological intelligence as inherently exhibiting the brittle, mechanical, and unsubtle properties that we have traditionally associated with machines. However, in this paragraph, Ray describes the essential process that I have proposed. Many contemporary projects have shown the feasibility of developing and expressing a deep understanding of "natural neural processes," and the language for that expression is mathematics.

We clearly will need to invent new computing devices to create the necessary capacity, and Ray appears to accept that this will happen. As for inventing new processes, new algorithms are continually being developed as well; however, what we have found thus far is that we have encountered few difficulties instantiating these models once they are revealed. The mathematical models derived from the reverse engineering process are readily implemented using available methods.

The revealed secrets of human intelligence will undoubtedly provide many enabling methods in the creation of the software of intelligence. An added bonus will be deep insight into our own nature, into human function and dysfunction.

Thomas Ray states that:

> The structure and function of the brain or its components cannot be separated. The circulatory system provides life support for the brain, but it also delivers hormones that are an integral part of the chemical information processing function of the brain. The membrane of a neuron is a structural feature defining the limits and integrity of a neuron, but it is also the surface along which depolarization propagates signals. The structural and life-support functions cannot be separated from the handling of information.

Ray goes on to describe several of the "broad spectrum of chemical communication mechanisms" that the brain exhibits. However, all of these are readily modelable, and a great deal of progress has already been made in this endeavor. The intermediate language is mathematics, and translating the mathematical models into equivalent nonbiological mechanisms is the easiest step in this process. With regard to the delivery of hormones by the circulatory system, this is an extremely low bandwidth phenomenon, which will not be difficult to model and replicate. The blood levels of specific hormones and other chemicals influence parameter levels that affect a great many synapses at once.

Ray concludes that "a metallic computation system operates on fundamentally different dynamic properties and could never precisely and exactly 'copy' the function of a brain." If one follows closely the progress in the related fields of neurobiology, brain scanning, neuron and neural region modeling, neuron-electronic communication, neural implants, and related endeavors, we find that our ability to replicate the salient functionality of biological information processing can meet any desired level of precision. In other words, the copied functionality can be "close enough" for any conceivable purpose or goal, including satisfying a Turing Test judge. Moreover, we find that efficient implementations of the mathematical models require substantially less computational capacity than the theoretical potential of the biological neuron clusters being modeled.

Ray goes on to describe his own creative proposal for creating nonbiological intelligence, which is to use evolutionary algorithms to allow a digital intelligence to "emerge," one that is "rooted in the nature of the medium." This would be, according to Ray, a "non-Turing intelligence," but "one which would complement rather than duplicate our talents and abilities."

I have no problem with this particular idea. Another of Ray's mistaken claims is that I offer human brain reverse engineering as the only route to strong AI. The truth is that I strongly advocate multiple approaches. I describe the reverse engineering idea (among others) because it serves as a useful existence proof of the feasibility of understanding and replicating human intelligence. In my book *The Age of Spiritual Machines*, I describe a number of other approaches, including the one that Ray prefers (evolutionary algorithms). My own work in pattern recognition and other aspects of AI consistently utilizes multiple approaches, and it is inevitable that the ultimate path to strong AI will combine insights from a variety of paradigms. The primary role of developing mathematical models of biological neurons, scanning the human brain, and reverse engineering the hundreds of brain regions is to develop biologically-inspired models of intelligence, the insights of which we will then combine with other lines of attack.

With regard to Ray's own preference for evolutionary or genetic algorithms, he misstates the scope of the problem. He suggests that the thousands of bits of genetic information in contemporary genetic algorithms may end up falling "ten orders of magnitude below organic evolution." But the thousands of bits of genetic order represented by contemporary systems are already only four or five orders of magnitude below that of the unique information contained in the genome.

Ray makes several other curious statements. Ray says:

> ...the most complex of our creations are showing alarming failure rates. Orbiting satellites and telescopes, space shuttles, interplanetary probes, the Pentium chip, computer operating systems, all seem to be pushing the limits of what we can effectively design and build through conventional approaches... Our most complex software (operating systems and telecommunications control systems) already contains tens of millions of lines of code. At present it seems unlikely that we can produce and manage software with hundreds of millions or billions of lines of code.

First of all, what alarming failure rates is Ray referring to? Computerized mission critical systems are remarkably reliable. Computerized systems of significant sophistication routinely fly and land our airplanes automatically. I am not aware of any airplane crash that has been attributed to a failure of these systems, yet many crashes are caused by the human errors of pilots and maintenance crews. Automated Intensive Care monitoring systems in hospitals almost never malfunction, yet hundreds of thousands of people die from human medical errors. If there are alarming failure rates to worry about, it's human failure, not those of mission critical computer systems. The Pentium chip problem which Ray alludes to was extremely subtle, caused almost no repercussions, and was quickly rectified.

The complexity of computerized systems has indeed been scaling up exponentially. Moreover, the cutting edge of our efforts to emulate human intelligence will utilize the self-organizing paradigms that we find in the human brain. I am not suggesting that self-organizing methods such as neural nets and evolutionary algorithms are simple or automatic to use, but they represent powerful tools which will help to alleviate the need for unmanageable levels of complexity.

Most importantly, it is not the case that the human brain represents a complexity comparable to "billions of lines of code." The human brain is created from a genome of only about 23 million bytes of unique information (less than Microsoft Word). It is through self-organizing processes that incorporate significant elements of randomness (as well as exposure to the real world) that this small amount of design information is expanded to the trillions of bytes of information represented in a mature human brain. The design of the human brain is not based on billions of lines of code or the equivalent thereof. Similarly, the task of creating human level intelligence in a nonbiological entity will not involve creating a massive expert system comprising billions of rules or lines of code, but rather a learning, chaotic, self-organizing, system, one ultimately that is biologically inspired.

Thomas Ray writes:

> The engineers among us might propose nano-molecular devices with fullerene switches, or even DNA-like computers. But I am sure they would never think of neurons. Neurons are astronomically large structures compared to the molecules we are starting with.

This is exactly my own point. The purpose of reverse engineering the human brain is not to copy the digestive or other unwieldy processes of biological neurons, but rather to understand their salient information processing methods. The feasibility of doing this has already been demonstrated in dozens of contemporary projects. The scale and complexity of the neuron clusters being emulated is scaling up by orders of magnitude, along with all of our other technological capabilities.

Once we have completed the reverse engineering of the several hundred brain regions, we will implement these methods in nonbiological substrates. In this way, we will combine these human-like capacities with the natural advantages of machines, i.e., the speed, accuracy and scale of memory, and, most importantly, the ability to instantly share knowledge.

Ray writes:

> Over and over again, in a variety of ways, we are shaping cyberspace in the form of the 3D material space that we inhabit. But cyberspace is not a material space and it is not inherently 3D. The idea of downloading the human mind into a computer is yet another example of failing to understand and work with the properties of the medium....Cyberspace is not a 3D Euclidean space. It is not a material world. We are not constrained by the same laws of physics, unless we impose them upon ourselves.

The reality is that we can do both. At times we will want to impose upon our virtual reality environments the three-dimensional gravity-bound reality we are used to. After all, that is the nature of the world we are comfortable in. At other times, we may wish to explore environments that have no earthly counterpart, ones that may indeed violate the laws of physics.

We can also do both with regard to emulating intelligence in our machines. We can apply Ray's preferred genetic algorithm approach while we also benefit from the reverse engineering of biological information processes, among other methods.

In summing up, Thomas Ray writes:

> Everything we know about intelligence is based on one example of intelligence, namely, human intelligence. This limited experience burdens us with preconceptions and limits our imaginations.

Actually, it is Thomas Ray who is limiting his imagination to his single idea of unleashing "evolution in the digital medium." Certainly there will be new forms of intelligence as nonbiological intelligence continues to grow. It will draw upon many sources of knowledge, some biologically motivated, and some inspired by our own imagination and creativity, ultimately augmented by the creativity of our machines.

The power of our civilization has already been greatly augmented by our technology, with which we are becoming increasingly intimate, as devices slip into our pockets and dangle from our bodies like jewelry. Within this decade, computing and communications will appear to disappear, with electronics being woven into our clothing, images being written directly to our retinas, and extremely high bandwidth wireless communication continually augmenting our visual and auditory reality. Within a few decades, with technology that travels inside our bodies and brains, we will be in a position to vastly expand our intelligence, our experiences, and our capacity to experience. Nonbiological intelligence will ultimately combine the inherent speed and knowledge sharing advantages of machines with the deep and subtle powers of the massively parallel, pattern recognition-based, biological paradigm.

Today, our most sophisticated machines are still millions of times simpler than human intelligence. Similarly, the total capacity of all the computers in the world today remains at least six orders of magnitude below that of all human brains, which I estimate at 10^{26} digitally controlled analog transactions per second. However, our biological capacity is fixed. Each of us is limited to a mere hundred trillion interneuronal connections. Machine intelligence, on the other hand, is growing exponentially in capacity, complexity, and capability. By the middle of this century, it will be nonbiological intelligence, representing an intimate panoply of paradigms, that predominates.

10

The Material World:
"Is That All There Is?"
Response to George Gilder
and Jay Richards

Ray Kurzweil

12.2.04

I n their foreword, George Gilder and Jay Richards describe me
(as well as John Searle and Thomas Ray) as "philosophical materialists," a term they define by quoting Carl Sagan's view of
the material world as "all there is, or ever was, or ever will be."
Kurzweil, Searle, and Ray, according to Gilder and Richards, "agree
that everything can or at least should be described in terms of chance
and impersonal natural law without reference to any sort of transcendent intelligence or mind. To them, ideas are epiphenomena of matter."

There are many concepts here to respond to. But my overriding reaction is: What's the problem with the so-called material world? Is the world of matter and energy not profound enough? Is it truly necessary to look beyond the world we encounter to find transcendence?

Where shall I start? How about water? It's simple enough, but consider the diverse and beautiful ways it manifests itself: the endlessly varying patterns as it cascades past rocks in a stream, then surging chaotically down a waterfall (all viewable from my office window, incidentally); the billowing patterns of clouds in the sky; the arrangement of snow on a mountain; the satisfying design of a single snowflake. Or consider Einstein's description of the entangled order and disorder in, well, a glass of water (i.e., his thesis on Brownian motion).

As we move into the biological world, consider the intricate dance of spirals of DNA during mitosis. How about the "loveliness" of a tree as it bends in the wind and its leaves churn in a tangled dance? Or the bustling world we see in a microscope? There's transcendence everywhere.

A comment on the word "transcendence" is in order here. To transcend means to "go beyond," but this need not compel us to an ornate dualist view that regards transcendent levels of reality (e.g., the spiritual level) to be not of this world. We can "go beyond" the "ordinary" powers of the material world through the power of patterns. Rather than a materialist, I would prefer to consider myself a "patternist." It's through the emergent powers of the pattern that we transcend.

Consider the author of this chapter. I am not merely or even principally the material stuff I am made of because the actual particles that comprise me turn over quickly. Most cells in the body are replaced within a few months. Although neurons persist longer, the actual atoms making up the neurons are also rapidly replaced. In the first chapter I made the analogy to water in a stream rushing around rocks. The pattern of water is relatively stable, yet the specific water molecules change in milliseconds. The same holds true for us human beings. It is the immense, indeed transcendent, power of our pattern that persists.

The power of patterns to persist goes beyond explicitly self-replicating systems such as organisms and self-replicating technology. It is the persistence and power of patterns that, quite literally, gives life to the Universe. The pattern is far more important than the material stuff that comprises it.

Random strokes on a canvas are just paint. But when arranged in just the right way, it transcends the material stuff and becomes art. Random notes are just sounds. Sequenced in an "inspired" way, we have music. A pile of components is just an inventory. Ordered in an innovative manner, and perhaps with some software (another pattern), we have the "magic" (i.e., transcendence) of technology.

We can regard the spiritual level as the ultimate in transcendence. In my view, it incorporates all of these, the creations of the natural world such as ourselves, as well as our own creations in the form of human technology, culture, art, and spiritual expression.

Is the world of patterns impersonal? Consider evolution. The "chance...impersonal" swirl of dust and wind gave rise to ever more intelligent, knowledgeable, creative, beautiful, and loving entities, and has done so at an ever accelerating pace. I don't regard this as an "impersonal" process because I don't regard the world and all of its attendant mysteries as impersonal. Consider what I wrote in the first chapter, that "technology is evolution by other means." In other words, technology is a continuation of the evolutionary process that gave rise to the technology creating species in the first place. It is another paradigm shift, a profound one to be sure, changing the focus from DNA-guided evolution to an evolutionary process directed by one its own creations, another level of indirection if you will.

If we put key milestones of both biological and human cultural-technological evolution on a single graph, in which the x-axis (number of years ago) and the y-axis (the paradigm shift time) are both plotted on exponential scales, we find a straight line with biological evolution leading directly to human-directed evolution.

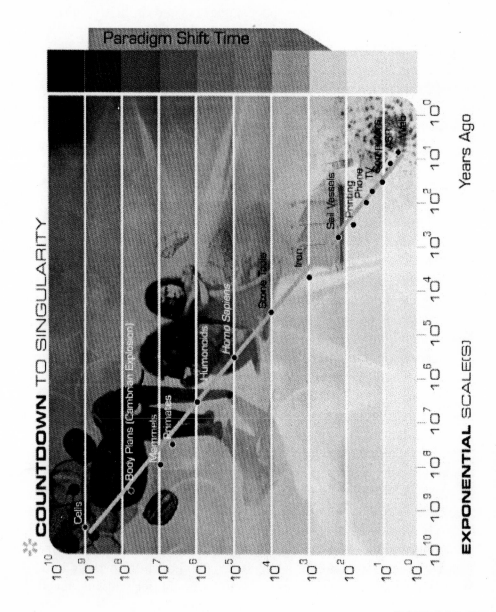

There are many implications of the observation that technology is an evolutionary process, indeed the continuation of the evolutionary process that gave rise to it. It implies that the evolution of technology, like that of biology, accelerates.

It also implies that technology, which is the second half of the evolutionary line above, and the cutting edge of evolution today, is anything but impersonal. Rather, it is the intensely human drama of human competition and innovation that George Gilder writes about (and makes predictions about) so brilliantly.

How about the first half of the line, the story of evolution that started with the swirling dust and water on an obscure planet? The personalness of the biological stage of evolution depends on how we view consciousness. My view is that consciousness, the seat of "personalness," is the ultimate reality, and is also scientifically impenetrable. In other words, there is no scientific test one can postulate that would definitively prove its existence in another entity. We assume that other biological human persons, at least those who are at least acting conscious, are indeed conscious. But this too is an assumption, and this shared human consensus breaks down when we go beyond human experience (e.g., the debate on animal consciousness, and by extension animal rights).

We have no consciousness detector, and any such device that we can imagine proposing will have built in assumptions about which we can debate endlessly. It comes down to the essential difference between objective (i.e., scientific) and subjective (i.e., conscious, personal) reality. Some philosophers then go on to say that because the ultimate issue of consciousness is not a scientific issue (albeit that the more superficial, i.e., the "easy" issues of consciousness as the philosopher David Chalmers describes them, can be amenable to scientific exploration), consciousness is, therefore, an illusion, or at least not a real issue. However, a more reasonable conclusion that one can come to, and indeed my own view, is that precisely because these central issues of reality are not fully resolvable by scientific experiment and argument alone, there is a salient role for philosophy and religion. However, this does not require a world outside the physical world we experience.

The arguments that I do make with regard to consciousness are for the sole purpose of illustrating the vexing and paradoxical (and in my view, therefore, profound) nature of consciousness, how one set of assumptions (i.e., that a copy of my mind file either shares or does not share my consciousness) leads ultimately to an opposite view, and vice versa.

So we could say that the universe—"all that is"—is indeed personal, is "conscious" in some way that we cannot fully comprehend. This is no more unreasonable an assumption or belief than believing that another person is conscious. Personally, I do feel this to be the case. But this does not require me to go beyond the "mere" "material" world and its transcendent patterns. The world that is, is profound enough.

Another conclusion that I come to in considering the acceleration of evolution is that ultimately the matter and energy in our vicinity will become infused with the intelligence, knowledge, creativity, beauty, and love of our human-machine civilization. And then our civilization will expand outwardly turning all the "dumb" matter we encounter into transcendently intelligent matter. So even in its present largely dumb state, the Universe has the potential for this explosion of intelligence, creativity, and other qualities we attribute to the spiritual aspect of reality. And if you do the math, because of the power of exponential growth, it won't take that long (mere centuries) to transform the Universe into smart matter. That is, of course, if we can figure out some way around the speed of light. I have some theories on this too, but I'll leave that train of thought for another time.

Let's consider the opposite direction for a moment, plumbing to the very smallest grains of reality. One would think that as we probe smaller and smaller aspects of the world, that they would become simpler and easier to understand. Yet, we've found the opposite to be the case. At the physically large level of reality that we live in, we often find a predictable Newtonian world, or at least we find many mechanisms that appear to work this way. Yet, as we consider the reality of a single photon, we encounter deep mysteries. We discover the photon simultaneously taking all paths available to it, only retroactively resolving the ambiguities in its path. We notice the photon taking on the properties of both wave and particle, an apparent mathematical contradiction. Photons are part of the material world, and we've got trillions of trillions of them. Is the world of matter and patterns not profound enough?

The Bible says that "the eyes of the Lord are in every place" (Proverbs 15:3). Admittedly, this view commonly expressed in both the Old and New Testaments is subject to dualist interpretations, but many religious traditions speak of God being manifest in the world as opposed to being a world apart. Dostoevsky (in *The Brothers Karamazov*) wrote, "Every blade of grass, every insect, ant, and golden bee, all . . . bear witness to the mystery of God and continually accomplish it themselves." Spinoza expressed it well when he wrote, "God reveals himself in the harmony of what exists."

The problem with conveying transcendent ideas with the models expressible in human language is that our language necessarily reduces their grandeur and subtlety. If we say that "God is everywhere," someone might interpret that to mean that God is some kind of gas that fills up the space, or an ether in which particles and waves move about. Language can only provide imperfect metaphors for transcendent thoughts. Thus apparently contradictory notions (i.e., God is manifest in the material world; versus God is everywhere but nonetheless of a different nature than the material world; versus God is an all powerful person with whom we can communicate and establish covenants; versus God created the world and is now watching from afar. . .) can be different views of the same transcendent reality. The ability to overcome apparent contradiction is what makes the spiritual aspect of reality transcendent. We also note with regret that these apparent contradictions are the source of much conflict.

Gilder and Richards' characterizations of Searle and Ray as philosophical materialists is fair enough, only theirs is a materialism stripped of any sense that the issue of consciousness introduces any mystery into our investigations. There is nothing special about Searle's concept of consciousness. Searle states that "consciousness is a biological process like digestion, lactation, photosynthesis." Searle goes on to say that "the brain is a machine, a biological machine to be sure, but a machine all the same. So the first step is to figure out how the brain does it and then build an artificial machine that has an equally effective mechanism for causing consciousness." Searle's view of consciousness is quite straightforward, and no different from, well, digestion. It is ironic that Searle has a reputation, totally undeserved in my view, for defending the deep mystery of the issue of consciousness, and the ultimate limits of scientific experimentation to resolve the issue.

As for Thomas Ray, consciousness hardly seems to impinge at all. Indeed, if we limit ourselves to scientific observation only (i.e., to objective, and therefore not subjective consideration), we can safely ignore it. Ray's view of quantum mechanics recognizes no difference between measurement and observation, because after all we need not concern ourselves with the who who is observing.

Denton describes his "awe" at the "eerie, other-worldly. . . impression" of the asymmetric patterns in nature. He contrasts the "self-organizing . . . self-referential . . . self-replicating . . . reciprocal . . . self-formative, and . . . holistic" qualities of designs in nature to the modular mechanisms of most contemporary machines. While I share Denton's sense of "wonderment" at the design principles manifest in nature, he makes the unsupported leap that patterns with such properties are inherently limited to biological processes, that machines could never display the same "other-worldly" (i.e., spiritual) dimension. I'd have to say that Denton does a good job of describing how transcendent attributes can be emergent properties of very complex systems. However, aside from pointing out the obvious limitations of much of contemporary technology, he fails to cite any compelling reason that our own creations are intrinsically restricted from emulating these powerful natural design principles.

As for Dembski, Gilder and Richards accurately describe his view as theistic, in the sense of an uncomfortable duality, with God and the spirit (i.e., consciousness) operating outside the material world. Many philosophers have pointed out the pitfalls of the dualistic view.

If God and spirit operate outside the material world and have no effect on it, then perhaps we can safely ignore them altogether. On the other hand, if they do affect and interact with the material world, then why not consider them part of it? Otherwise, our metaphysics becomes hopelessly elaborate.

Gilder and Richards describe my view as "a substitute vision for those who have lost faith in the traditional object of religious belief." The traditional object of religious belief is often referred to as God. But if we are to understand God as infinite in intelligence, knowledge, creativity, and so on, then it would seem reasonable to explore new metaphors to attempt to express what is inherently not fully expressible in our finite language. To restrict our view of God to only one tradition limits Who should be regarded as without limit. The words and stories of our ancient traditions may indeed lose their resonance over time, not because the timeless truths have changed, and not because of any inconsistency with our expanding scientific knowledge, but rather because they were attempts to express transcendent ideas in language poorly equipped for such a purpose. It makes sense to update not the truths themselves but our expressions of these truths in keeping with our evolving understanding of the world we live in.

Furthermore, it is not my view that "the very notion of improvement" is "alien in a materialistic universe." One of the ways in which this universe of evolving patterns of matter and energy that we live in expresses its transcendent nature is in the exponential growth of the spiritual values we attribute in abundance to God: knowledge, intelligence, creativity, beauty, and love.

Joy Drives Off the Road?

Fundamentally, Gilder and Richards and I share a deeply critical reaction to Bill Joy's prescription of relinquishment of "our pursuit of certain types of knowledge." Just as George Soros attracted attention by criticizing the capitalist system of which he was a primary beneficiary, the credibility of Joy's treatise on the dangers of future technology has been enhanced by his reputation as a primary archi-

tect of contemporary technology. Being a technologist, Joy claims
not to be anti-technology, saying that we should keep the beneficial
technologies, and relinquish only those dangerous ones, like
nanotechnology. The problem with Joy's view is that the dangerous
technologies are exactly the same as the beneficial ones. The same
biotechnology tools and knowledge that will save millions of future
lives from cancer and other diseases could potentially provide a terrorist with the means for creating a bioengineered pathogen. The
same nanotechnology that will eventually help clean up the environment and provide material products at almost no cost are the same
technologies that could be misused to introduce new nonbiological
pathogens.

I call this the deeply intertwined promise and peril of technology,
and it's not a new story. Technology empowers both our creative
and destructive natures. Stalin's tanks and Hitler's trains used technology. Yet few people today would really want to go back to the
short (human lifespan less than half of today's), brutish, disease-filled,
poverty-stricken, labor-intensive, disaster-prone lives that 99 percent
of the human race struggled through a few centuries ago.

We can't have the benefits without at least the potential dangers.
The only way to avoid the dangerous technologies would be to relinquish essentially all of technology. And the only way to accomplish
that would be a totalitarian system (e.g., Brave New World) in which
the state has exclusive use of technology to prevent everyone else
from advancing it. Joy's recommendation does not go that far obviously, but his call for relinquishing broad areas of the pursuit of knowledge is based on an unrealistic assumption that we can parse safe and
risky areas of knowledge.

Another reason that Joy's call for relinquishment of broad areas
such as nanotechnology is unrealistic is that nanotechnology is not a
simple unified field. Rather, it is the inevitable end result of the
ongoing exponential trend of miniaturization in all areas of technology, which continues to move forward on hundreds of fronts (we're
currently shrinking both electronic and mechanical technology by a
factor of 5.6 per linear dimension per decade). It's not feasible to

stop nanotechnology or other broad areas of technology without stopping virtually all technology.

In an article on this same issue, titled "Stop everything . . . It's Techno-Horror!" in the March 2001 issue of *The American Spectator*, George Gilder and Richard Vigilante write, "in the event of . . . an unplanned bio-catastrophe, we would be far better off with a powerful and multifarious biotech industry with long and diverse experience in handling such perils, constraining them, and inventing remedies than if we had 'relinquished' these technologies to a small elite of government scientists, their work closely classified and shrouded in secrecy."

I agree quite heartily with this eloquent perspective. Consider as a contemporary test case, how we have dealt with one recent technological challenge. There exists today a new form of fully nonbiological self-replicating entity that didn't exist just a few decades ago: the computer virus. When this form of destructive intruder first appeared, strong concerns were voiced that as they became more sophisticated, software pathogens had the potential to destroy the computer network medium they live in. Yet the "immune system" that has evolved in response to this challenge has been largely effective. Although destructive self-replicating software entities do cause damage from time to time, the injury is but a tiny fraction of the benefit we receive from the computers and communication links that harbor them.

One might counter that computer viruses do not have the lethal potential of biological viruses or of destructive future nanotechnology. Although true, this only strengthens my observation. The fact that computer viruses are not usually deadly to humans (although they can be if they intrude on mission critical systems such as airplanes and intensive care units) only means that more people are willing to create and release them. It also means that our response to the danger is relatively relaxed. Conversely, when it comes to future self-replicating entities that may be potentially lethal on a large scale, our response on all levels will be vastly more intense.

Joy's treatise is effective because he paints a picture of future dangers as if they were released on today's unprepared world. The reality is that the sophistication and power of our defensive technologies and knowledge will grow along with the dangers. When we have gray goo, we will also have blue goo ("police" nanobots that combat the "bad" nanobots). The story of the twenty-first century has not yet been written, so we cannot say with assurance that we will successfully avoid all misuse. But the surest way to prevent the development of the defensive technologies would be to relinquish the pursuit of knowledge in broad areas, which would only drive these efforts underground where they would be dominated by the least reliable practitioners (e.g., the terrorists).

There is still a great deal of suffering in the world. Are we going to tell the millions of cancer patients that we're canceling all cancer research despite very promising emerging treatments because the same technology might be abused by a terrorist? Consider the following tongue-in-cheek announcement, which I read during a radio debate with Joy: "Sun Microsystems announced today that it was relinquishing all research and development that might improve the intelligence of its software, the computational power of its computers, or the effectiveness of its networks due to concerns that the inevitable result of progress in these fields may lead to profound and irreversible dangers to the environment and even to the human race itself. 'Better to be safe than sorry,' Sun's Chief Scientist Bill Joy was quoted as saying. Trading of Sun shares was automatically halted in accordance with Nasdaq trading rules after dropping by 90 percent in the first hour of trading." Joy did not find my mock announcement amusing, but my point is a serious one: Advancement in a broad array of technologies is an economic imperative.

Although I agree with Gilder and Vigilante's opposition to the essentially totalitarian nature of the call for relinquishment of broad areas of the pursuit of knowledge and technology, their *American Spectator* article directs a significant portion of its argument against the technical feasibility of the future dangers. This is not the best strategy in my view to counter Joy's thesis. We don't have to look further than today to see that technology is a double-edged sword. Gilder has written with great enthusiasm and insight in his books and newsletters of the exponential growth of many technologies, including Gilder's Law on the explosion of bandwidth. In my own writings, including in this book, I have shown how the exponential growth of the power of technology is pervasive and affects a great multiplicity of areas. The impact of these interacting and accelerating revolutions is significant in the short-term (i.e., over years), but revolutionary in the long term (i.e., over decades). I believe that the most cogent strategy to oppose the allure of the suppression of the pursuit of knowledge is not to deny the potential dangers of future technology nor the theoretical feasibility of disastrous scenarios, but rather to build the case that the continued relatively open pursuit of knowledge is the most reliable (albeit not foolproof) way to reap the promise while avoiding the peril of profound twenty-first century technologies.

I believe that George and I are in essential agreement on this issue. In the *American Spectator* article, he and Richard Vigilante write the following which persuasively articulates the point:

"Part of the 'mysterious' realm that Einstein called 'the cradle of all true art and true science,' chance is beyond the ken of inductive reason. When Albert Hirschman writes that 'creativity always comes as a surprise to us,' he is acknowledging this essential property of invention. Any effort to reduce the world to the dimensions of our own present understanding will exclude novelty and progress. The domain of chance is our access to futurity and to providence. 'Trusting to chance' seems terrifying, but it is the only way to be open to possibility."

Printed in the United States
20559LVS00005B/114